MW00845481

LMA 790-1

LUNAR EXCURSION MODULE

FAMILIARIZATION MANUAL

GRUMMAN AIRCRAFT ENGINEERING CORPORATION

NAS 9-1100

EXHIBIT E, PARAGRAPH 10.2

TYPE 1 DOCUMENT, APPROVED BY NASA 20 DAY CLAUSE

THIS MANUAL SUPERSEDES LMA 790-1 DATED 15 JANUARY 1964

PUBLICATIONS GROUP/SERVICE AND PRODUCT SUPPORT DEPARTMENT/GRUMMAN AIRCRAFT ENGINEERING CORPORATION/BETHPAGE/NEW YORK

ISBN #978-1-935700-66-1
www.PeriscopeFilm.com

15 July 1964

LIST OF EFFECTIVE PAGES

INSERT LATEST CHANGED PAGES. DESTROY SUPERSEDED PAGES.

NOTE: The portion of the text affected by the changes is indicated by a vertical line in the outer margins of the page.

Manuals will be distributed as directed by the NASA Apollo Project Office. All requests for manuals should be directed to the NASA Apollo Spacecraft Project Office at Houston, Texas.

TABLE OF CONTENTS

TABLE OF CONTENTS (cont)

LIST OF ILLUSTRATIONS

LIST OF TABLES

FOREWORD

This Familiarization Manual provides an operational description of all subsystems and major components of the lunar landing LEM. The information contained herein is for orientation and indoctrination purposes only. The scope of coverage describes the LEM mission, spacecraft structure, operational subsystems, prelaunch operations, and ground support equipment.

SECTION I

MISSION DESCRIPTION

1-1. GENERAL

The Lunar Excursion Module (LEM) System consists of a manned vehicle (module) and various related subsystems. The LEM System enables successful completion of the LEM mission, utilizing the concept known as the Lunar Orbital Rendezvous (LOR) technique. The LEM mission which is part of the overall Apollo mission, begins shortly after separation of the LEM from the Command/Service modules continues through lunar descent, lunar stay, lunar ascent, and ends at rendezvous with the orbiting Command/Service module prior to the return to earth.

1-2. LEM MISSION.

Prior to earth launch, the LEM System is subjected to rigorous checkouts to assure complete mission reliability and maximum crew safety. System acceptance and functional tests, integrated equipment tests, assembly tests, launch pad tests, and the countdown operation permit constant system monitoring. During these tests, each subsystem is checked to that level possible without equipment removal. A general-purpose spacecraft-checkout system, the Acceptance Checkout Equipment-Spacecraft (ACE-S/C) is used for computer-controlled or manually controlled acceptance tests and prelaunch tests of the LEM system. The complete LEM system is exercised through its various modes of operation, redundancies are isolated and checked, and diagnostic routines are performed to the replaceable-unit level.

Prior to and at launch the landing gear of the LEM is folded for storage and the LEM is installed inside the adapter immediately above the S-IVB third stage. The Command/Service modules (the Command module contains the three-man crew) are attached to the LEM and S-IVB by the adapter. This arrangement forms the nose of the Saturn. When launched, the Saturn vehicle injects the payload, consisting of the S-IVB stage, LEM and the Command/Service modules, into an earth orbit. During this orbit, the LEM is still attached to the apex of the third stage with legs folded and antenna retracted.

When this orbit has been achieved, the three astronauts in the Command module perform subsystem checkouts in preparation for escape from the earth orbit. At this time, certain in-flight maintenance operations may be performed, if necessary. Mission abort may be executed now or at any other time during the overall mission. While in orbit, the Command module performs landmark sightings and other navigation and guidance tasks. Upon completion of earth orbit (normally two revolutions), the S-IVB third stage engine is restarted and injects the LEM and Command/Service modules into an earth-moon trajectory (translunar injection).

After the initial coast of translunar flight, the Command/Service modules detach as a unit from the LEM and S-IVB stage, rotate 180° in free flight, and dock with the nose of the Command module on the top hatch of the LEM. This maneuver is called transposition. During transposition, the LEM and S-IVB stage are stabilized by the Stabilization and Control System of the S-IVB stage. Upon completion of transposition, the spent S-IVB stage and the lower part of the LEM adapter are jettisoned.

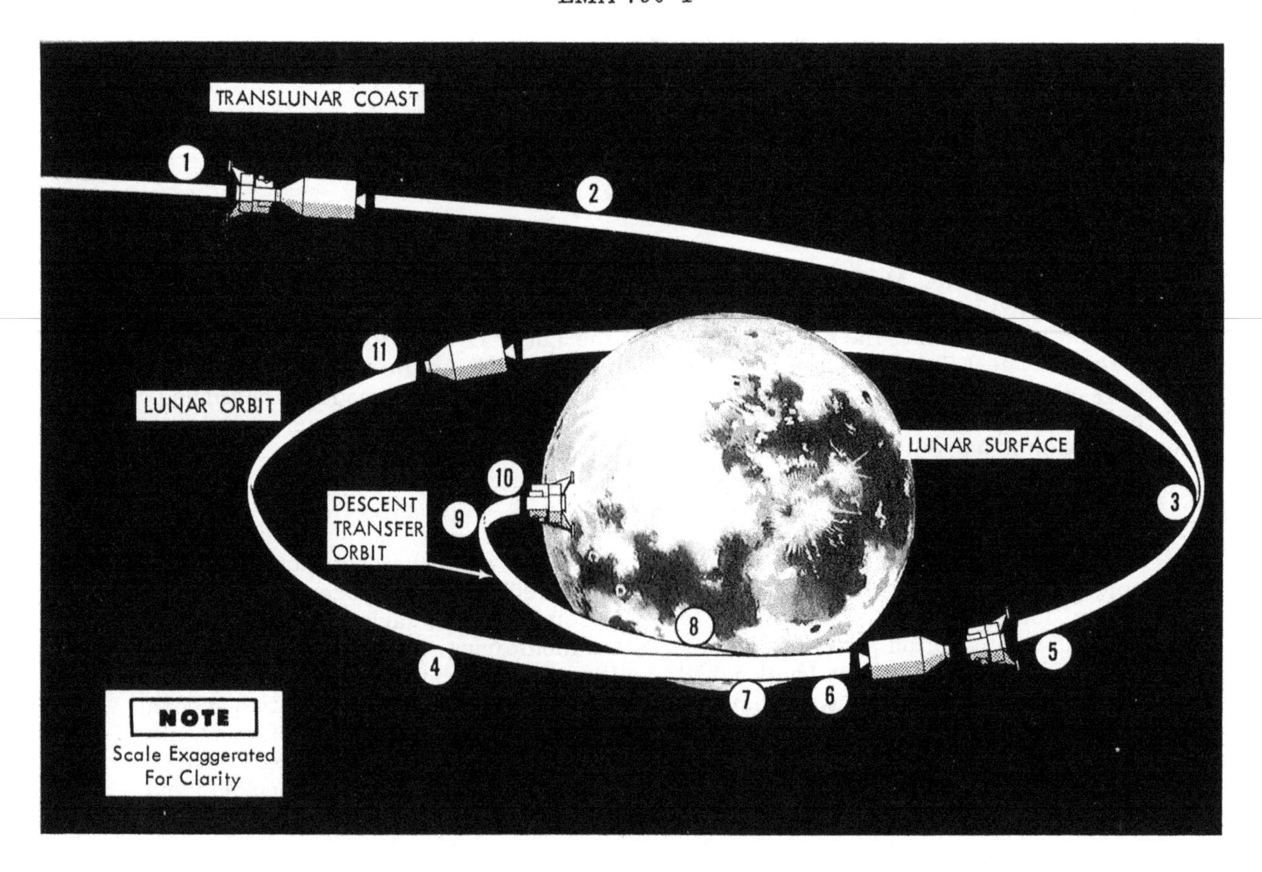

(1) Transposition near earth.

(2) LEM and Command/Service modules (C/SM) continue translunar coast. LEM remains passive.

(3) Service Module Propulsion System injects LEM-C/SM into 80 nmi lunar orbit.

(4) Two crew members enter LEM and perform lunar orbit checkout.

(5) LEM crew aligns Navigation & Guidance Subsystem prior to separation from C/SM.

(6) Reaction Control Subsystem separates LEM from C/SM.

(7) Descent engine injects LEM into descent transfer orbit. C/SM remains in lunar orbit.

(8) Descent engine cuts off and LEM coasts to pericynthion.

(9) Descent engine injects LEM into initial powered descent which ends with LEM situated 20 nmi from proposed landing site.

(10) LEM enters final powered descent to 1000 ft and proceeds to touchdown.

(11) C/SM remains in lunar orbit for eventual rendezvous with LEM after completion of lunar stay.

201LMA10-13-1

Figure 1-1. Mission Profile (Sheet 1)

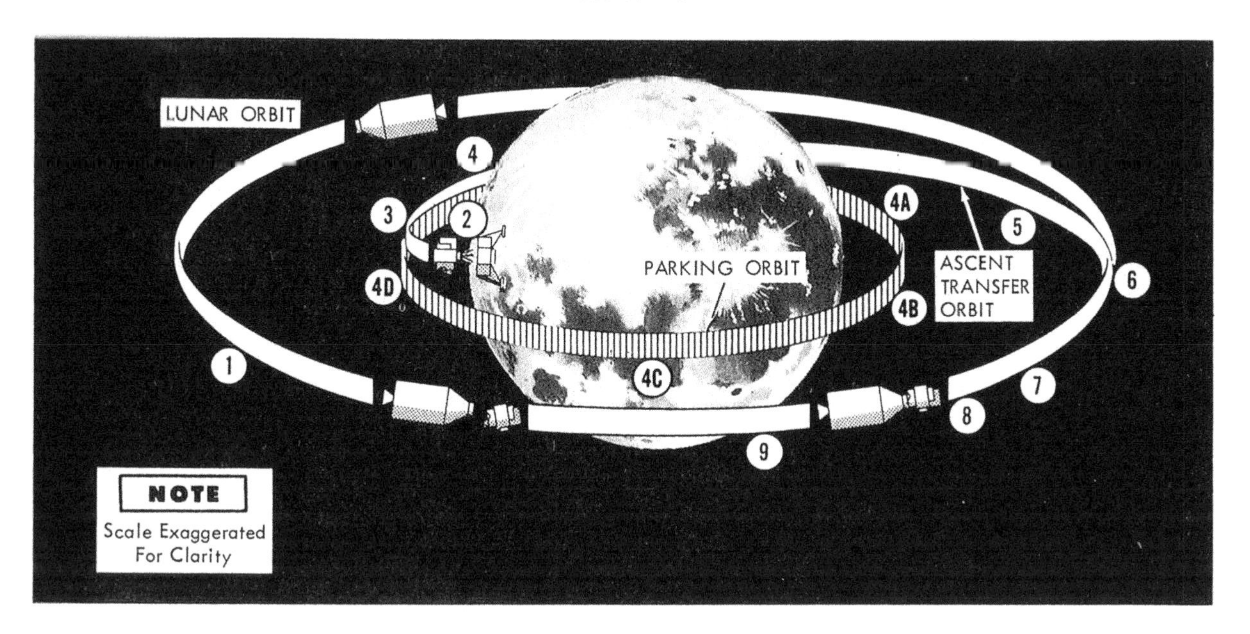

1. With C/SM in lunar orbit, proper LEM launch time occurs when C/SM has passed to a point slightly downrange of its zenith position overhead.

2. LEM ascent stage is separated from descent stage and ascent engine is fired for lunar launch. Descent stage remains on lunar surface.

3. LEM reaches pericynthion; ascent engine is cut off at 50,000 ft.

4. If LEM was launched at proper time, it enters directly into the ascent transfer orbit. If launch was delayed, LEM enters parking orbit.

   4A. Rendezvous radar and landmark sightings are used to determine LEM orbit and relative position of C/SM. Crew also determines time at which ascent engine must be fired to inject LEM into ascent transfer orbit.

   4B. LEM placed in minimum power consumption state; all unnecessary equipment is shut down.

   4C. Ninety minutes before the ascent engine must be re-started, LEM is prepared for injection into ascent transfer orbit; the Inertial Measuring Unit is aligned while the Rendezvous Radar tracks the C/SM.

   4D. Ascent engine is re-started; LEM now enters ascent transfer orbit.

5. Ascent transfer orbit carries LEM to a point within rendezvous range of C/SM.

6. LEM is now in rendezvous range (within 5 nmi from C/SM).

7. When LEM is approximately 500 feet from C/SM, LEM commander manually maneuvers LEM to a docking attitude and adjusts rate of closure until complete docking is accomplished.

8. LEM is secured to C/SM. Crew transfers to Command module.

9. LEM is jettisoned, concluding its mission. Service module engine is fired, taking C/SM out of lunar orbit and into earth trajectory.

2011MA10-13-2

Figure 1-1. Mission Profile (Sheet 2)

With the adapter jettisoned, the LEM landing gear becomes extended. When the jettisoned S-IVB stage is clear, the Command/Service modules and the LEM are oriented for the translunar coast period (figure 1-1). During the translunar coasting flight, the LEM remains passive, except for the Inertial Measurement Unit (IMU) heaters and portions of the Environmental Control Subsystem (ECS), Electrical Power Subsystem (EPS), and Instrumentation Subsystem. The Command/Service modules perform all navigational guidance functions and, implemented by the Service module reaction controls, initiate midcourse correction maneuvers.

Approximately 70 hours after launch, the Service module Propulsion System injects the LEM and Command/Service modules into an 80-nmi circular lunar orbit. During the first part of this lunar orbit, the astronauts in the Command module perform landmark sightings for orbit determination and Navigation and Guidance Subsystem updating. Upon completion of the first portion of the lunar orbit, the LEM is pressurized from the Command module and the LEM internal environment is checked. Two astronauts enter the LEM through the top docking hatch; the third astronaut remains in the Command module. The 2 man LEM crew then performs a checkout of each subsystem using the on board equipment. Upon completion of the checkout, the two astronauts align the Navigation and Guidance Subsystem. After a final landmark sighting by the Command module, the navigation and guidance parameters of the Command module are compared with those of the LEM to validate subsystem performance. The inertial altitude reference of the LEM Abort Guidance Section is then aligned with respect to the primary system. At a predetermined point in lunar orbit, the Reaction Control Subsystem (RCS) separates the LEM from the Command/Service modules. The astronauts prepare the LEM for descent by performing a visual inspection, followed by checkout of the transmission capability of the Communication Subsystem and the tracking capability of the rendezvous and landing radar.

The descent stage of the LEM Propulsion Subsystem injects the LEM into an elliptical Hohmann descent transfer orbit. This orbit has a pericynthion of 50,000 feet approximately 175 miles up range of the proposed landing site. At the conclusion of the injection thrust phase the descent engine cuts off and the LEM begins its coast toward the pericynthion. The LEM Stabilization and Control Subsystem operates in the minimum limit cycle mode during all main rocket engine firings.

During the initial part of the descent transfer orbit coast, the descent transfer orbit ephemeris is verified by using the Command module to track the LEM, or the LEM radar to track the Command module. At some point in this orbit, the astronauts may have to update the IMU of the Navigation and Guidance Subsystem by star sightings and lunar landmark sightings because more than an hour will have elapsed between alignments.

At the pericynthion of the descent transfer orbit, the descent engine of the Propulsion Subsystem is fired to initiate powered descent. Descent to the lunar surface consists of three distinct maneuvers: braking at approximately 50,000 to 10,000 feet, flare at approximately 10,000 to 200 feet, translation and terminal descent to touchdown. The braking maneuver is performed at near maximum descent engine thrust along a near optimum (minimum fuel) trajectory. The flare maneuver is performed at nearly constant flight path angle and vehicle attitude and the trajectory is shaped to allow visibility of the landing site. The final phase to touchdown provides the astronauts with the capability of close up landing site inspection and site selection. During powered descent, landing radar data and IMU information is used. Starting at an altitude of approximately 25,000 feet, the radar data is statistically mixed with the

IMU information. Because the inertial data degrades with time while the radar data improves, the radar information is weighted more heavily as the descent progresses, until, near touchdown radar derived altitude and velocity data may be used at a 100% weighting factor for IMU updating. Direct display of landing radar data for astronaut utilization during manual phases of the landing maneuver is also provided.

Under normal conditions, the descent engine is ignited when the LEM arrives at the pericynthion. The descent stage reduces the LEM's velocity during its descent to the lunar surface. The engine is gimballed so that the center of gravity lies along the thrust vector at all times. Descent of the LEM is performed under control of the N&GS to approximately 200 feet above the lunar surface. At this point, the LEM crew may take control of the vehicle to select the best landing site and perform the final descent to touchdown. To accomplish translation to a desired spot on the lunar surface the thrust vector may be tilted to accelerate the LEM in the direction of the landing site. At approximately 3 feet above the lunar surface, the engine is cut off and the LEM free falls to the lunar surface. An abort may be executed at any point up to and including the actual lunar touchdown, by means of powered ascent and rendezvous with the orbiting Command/Service modules.

After LEM touchdown on the lunar surface, the two astronauts perform a checkout of all subsystems to determine whether damage occurred upon landing and to assure that all systems are capable of performing the functions required for a successful ascent. The decision is then made as to whether the normal planned stay time operations can be executed. If no damage has occurred or if the damage is remedial, the crew then prepares for one astronaut to leave the LEM. The preparations consist of surveying the surrounding lunar landscape, checking the LEM hatches, and performing a final check of the portable life support systems (PLSS). All equipment not essential for lunar stay is turned off.

After the LEM is secured for lunar stay, it is depressurized and one astronaut leaves. The LEM is then pressurized. The exterior of the LEM is inspected by the exploring astronaut and the communication antennas are deployed. A television system is used to send pictures of the lunar scene back to earth via the S-band link. Photographic records are also made and samples of the lunar surface are collected. The astronaut outside the LEM is always in direct visual and voice contact (via VHF) with the astronaut inside the LEM. After approximately 3 hours of exploration, the astronaut must return to the LEM to replenish his life support pack. Life support stores permit three refills of each of the two portable life support packs.

Upon completion of lunar surface exploration, the LEM is depressurized and the exploring astronaut enters. After LEM pressurization, the portable life support system is replenished. For the PLSS to have a 4-hour operating capacity, a 6-hour battery recharge period is required. A voice report is made to earth via the S-band link and pertinent scientific data is transmitted and recorded. Subsequently, additional lunar surface exploration phases may occur depending upon the planned stay time. Up to four, four hour excursions are possible within the constraint of consumable provisioning.

When the lunar stay is completed, the astronauts prepare the LEM for launch and ascent. A complete checkout is made of all subsystems. The Navigation and Guidance Subsystem and the Abort Guidance Section of the Stabilization and Control Subsystem are aligned. The LEM location relative to the position of the lunar orbiting Command/Service modules is determined with the rendezvour radar. The Alignment Optical Telescope (AOT) obtains celestial data for alignment of the IMU.

Because the LEM descent stage is left behind at launching, all connections between the ascent and descent stages (including electrical cabling and oxygen piping) are severed at this time. The LEM is now ready for launch and ascent from the lunar surface and eventual rendezvous with the orbiting Command/Service modules. Nominal launch time occurs when the Command/Service modules in lunar orbit are slightly downrange of their zenith position over the LEM. Assuming the LEM is launched at this time, the ascent engine burns continuously from lift off to injection into an ascent transfer orbit (approximately 6 minutes). The ascent trajectory begins with a vertical rise, followed by two pitchover phases (one at a high pitch rate and a final one at a comparatively low pitch rate), with burnout occurring at 50,000 feet. At this point, the LEM is in an ascent transfer orbit, which nominally intersects the Command/Service modules orbit at conjunction.

If the launch of the LEM from the lunar surface is briefly delayed, a short stay in a parking orbit is required after reaching an altitude of 50,000 feet, to attain the proper relative position of the LEM and the Command/Service modules for injection into a rendezvous trajectory. If lunar surface launch takes place at the most unfavorable time, up to 18 hours of phasing time may be required to be spent in the parking orbit. All unnecessary equipment is turned off while in the parking orbit. Approximately 60 minutes before firing the ascent engine, the LEM is prepared for injection into the ascent transfer orbit. Preparation consists of aligning the IMU, and tracking the Command/Service modules with the rendezvous radar. The ascent engine is then fired at the appropriate time to place the LEM in the ascent transfer orbit.

When the LEM is in the free-flight ascent transfer orbit, it is carried from 50,000 feet to within rendezvous range of the Command/Service modules. Rendezvous begins at approximately 5 nmi and terminates approximately 500 feet from the lunar orbiting Command/Service modules. When the LEM is approximately 500 feet from the Command/Service modules, a LEM astronaut manually maneuvers the LEM to a docking attitude and adjusts the rate of closure until complete docking is accomplished. During the docking phase, the RCS thrusters are used in the manual modes of the Stabilization and Control Subsystem. The Command/Service modules normally remain passive during rendezvous, although they are also capable of maneuvering for rendezvous, if necessary.

After docking is completed, the LEM is secured to the Command/Service modules and the two astronauts in the LEM prepare for transfer to the Command module. Pressures are equalized, LEM subsystems are turned off, and scientific samples are transferred to the Command module. After the two astronauts transfer from the LEM through the docking hatch, to the Command module, the LEM is jettisoned. This concludes the LEM mission. The Service module engine is then fired to take the Command/Service modules out of lunar orbit and into an earth trajectory. The Service module is jettisoned before the Command module is oriented for reentry into the earth's atmosphere.

SECTION II

LEM STRUCTURE

2-1. GENERAL.

The LEM structure (figure 2-1) consists of two separate stages: the descent stage and the ascent stage. The stages are mated to form a self-sustaining structure, with the adjoining faces of the stages providing system continuity. Provision is made for separating the stages and the interconnecting umbilicals at lunar launch, or for abort at any time during the mission. The approximate dimensions of the LEM and a front, side, and top view are shown in figure 2-2. The Earth launch weight of the LEM is approximately 30,000 pounds.

2-2. ASCENT STAGE.

The ascent stage (figure 2-3) is the manned portion of the LEM. The crew is housed in the crew compartment, from where they control the flight, lunar landing, lunar launch, and rendezvous and docking with the Command/Service modules. The crew compartment is also used as the operations center for the crew during the lunar stay. In addition to the crew compartment, the ascent stage consists of the midsection, aft equipment bay, tankage sections, windows, and hatches. The ascent engine and related components are located at the vehicle's center of gravity in the midsection. Pressure and temperature within the crew compartment and midsection are controlled by the ECS. The aft equipment bay and tankage sections are not pressurized or temperature-controlled.

The ascent stage is constructed of aluminum alloy. The structural skin is surrounded with a layer of insulation and a very thin aluminum skin that provides thermal protection. The outer skin is approximately 2 inches from the structural skin. The crew compartment contains two triangular windows, the forward hatch and tunnel, controls and indicators, and items necessary for crew comfort and support. The midsection is a smaller cylindrical section, directly behind the crew compartment. The midsection contains the ascent engine hatch, top hatch, Environmental Control Subsystem (ECS), and stowage for equipment that must be accessible to the crew.

The upper docking tunnel, at the top centerline of the ascent stage, is used when transposition is performed and to transfer the crew during translunar flight. The forward docking tunnel, at the lower front of the crew compartment, is used for entering and leaving the LEM while on the lunar surface and for return of the crew to the Command module after departure from the lunar surface. Pressure-tight, plug-type hatches in each tunnel are manually opened and closed and sealed with formed-in-place silicone elastomeric seals.

The aft equipment bay is separated from the midsection by a pressure-tight bulkhead. This area houses the glycol loop and GOX accumulator for the ECS, two helium tanks for ascent engine pressurization, inverters, batteries, electrical equipment, two cryogenic hydrogen tanks and a cryogenic oxygen tank for the Electrical Power Subsystem (EPS). The propellant tankage sections are on either side of the midsection outside the pressurized area. The tankage sections contain the ascent engine fuel and oxidizer tanks; Reaction Control Subsystem fuel, oxidizer, and helium tanks;

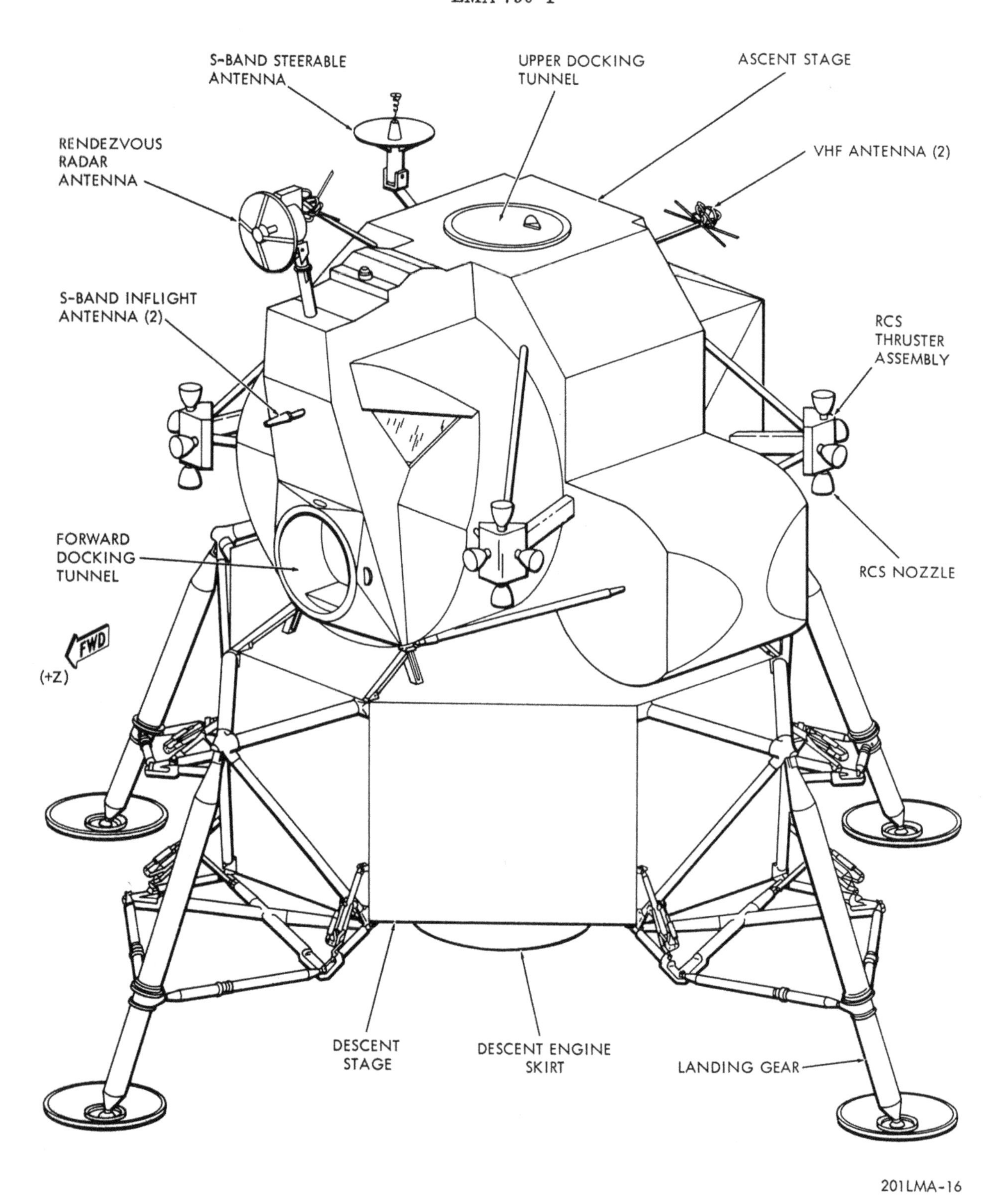

Figure 2-1. LEM Structure

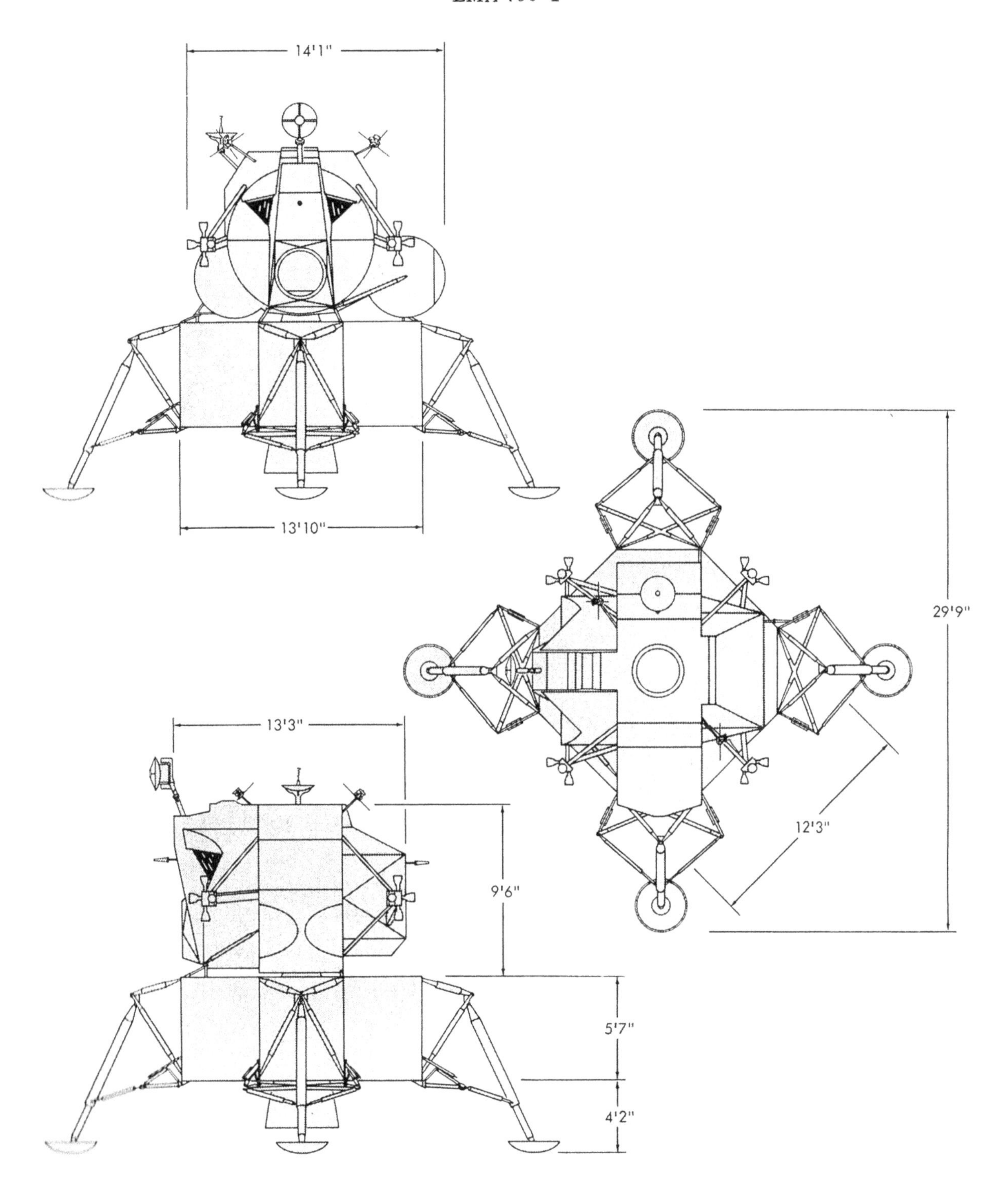

Figure 2-2. LEM Dimensions

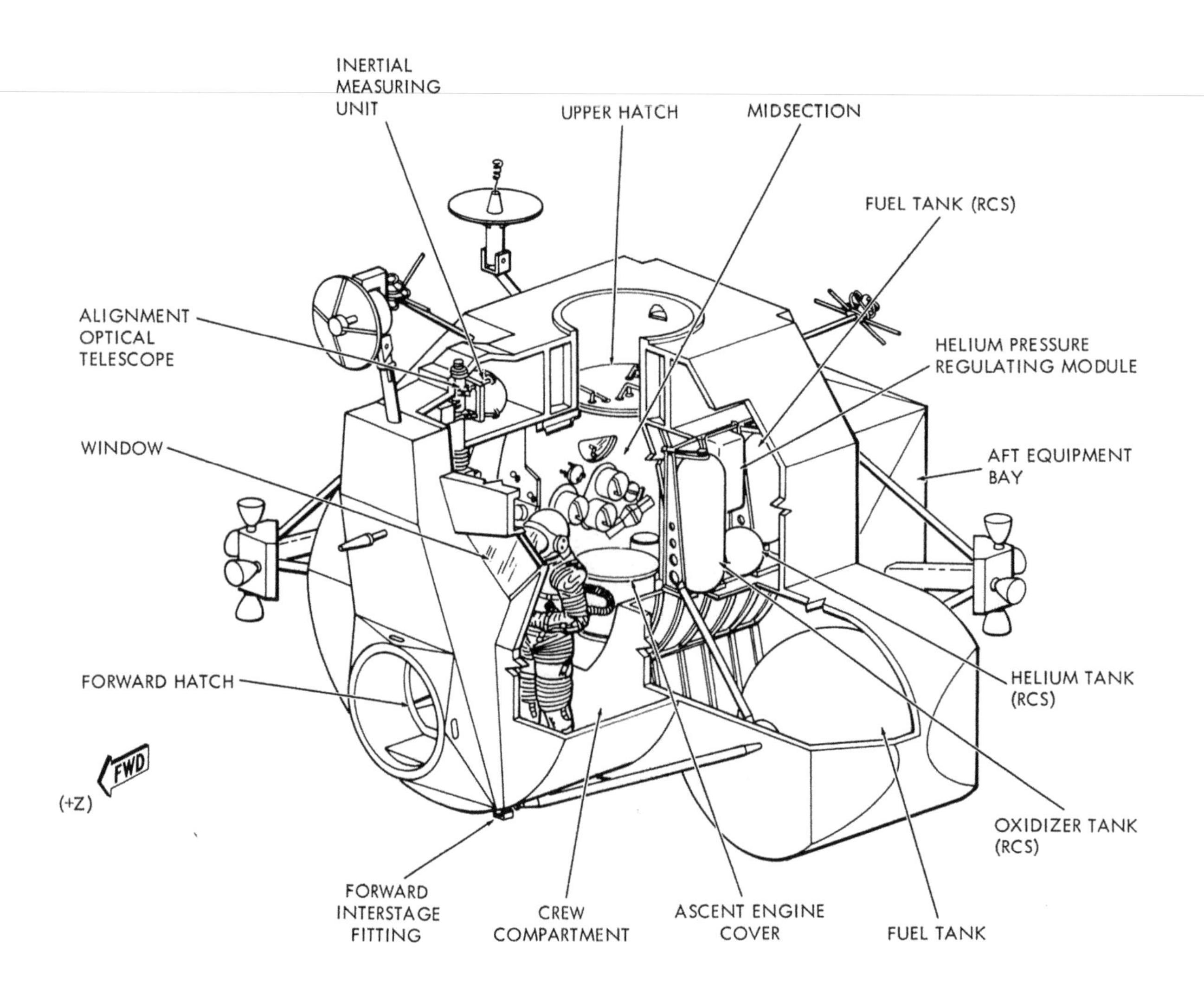

Figure 2-3. Ascent Stage (Sheet 1)

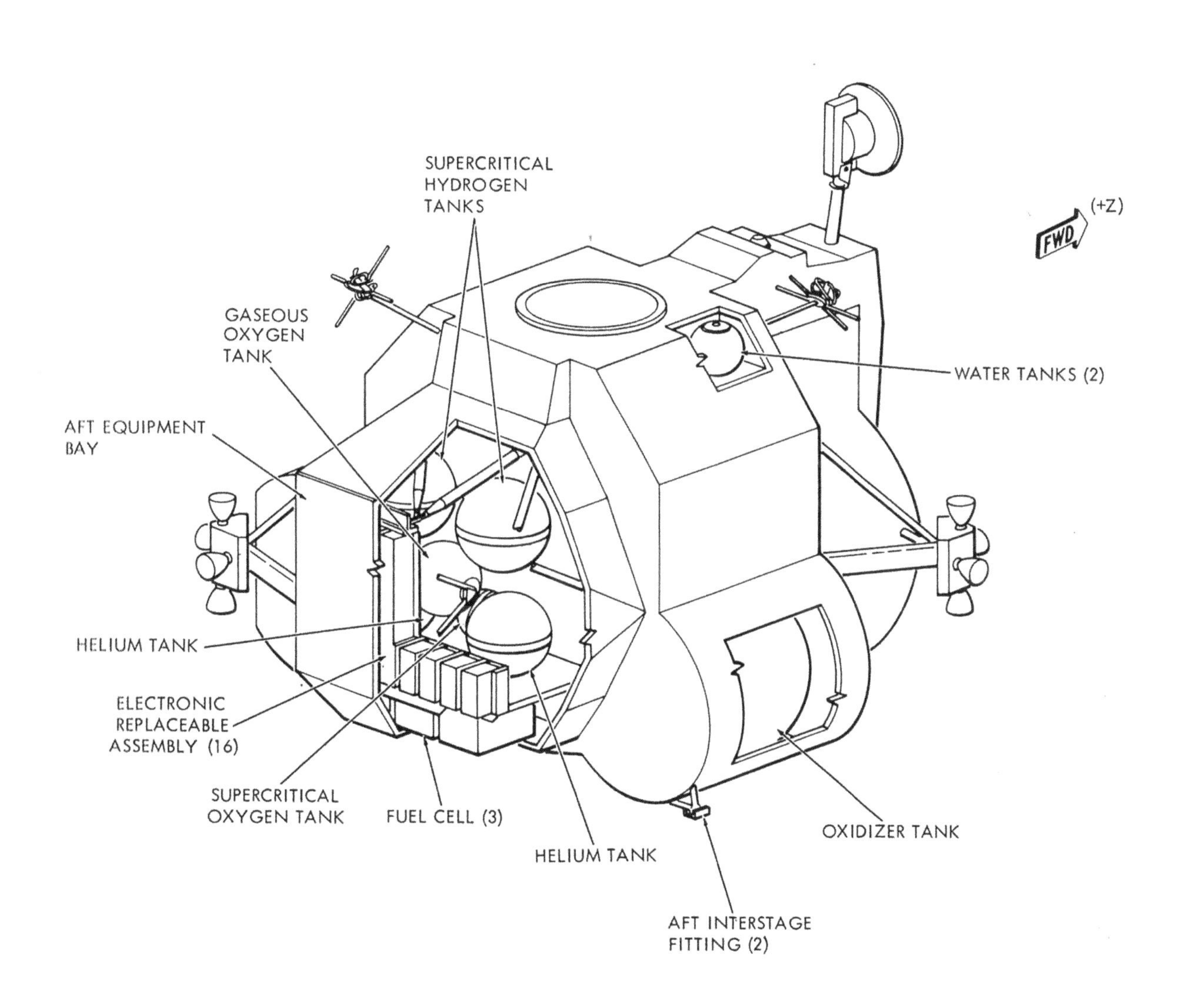

Figure 2-3. Ascent Stage (Sheet 2)

and ECS water tanks. The ratio of oxidizer to fuel is 1.6 to 1 by weight; therefore, the ascent engine propellant tanks are offset to one side to maintain the lateral center of gravity on the X-axis.

2-3. WINDOWS.

Two triangular windows in the front face of the crew compartment provide visibility during the descent transfer orbit, lunar landing, and the rendezvous and docking phases of the mission. Both windows have approximately 2 square feet of viewing area and are canted down and to the side to permit adequate sideways and downward visibility. Each window consists of two panes separated by a space. The outer pane is Chemcor glass; the inner pane, Stretchplex 55. A clamp-type seal that consists of a Teflon TFE jacket surrounding a metallic spring secures each pane.

2-4. HATCHES.

Two similar hatches in the ascent stage permit entering and leaving the LEM. The upper hatch is on the X-axis, directly above the ascent engine cover; the forward hatch is on the Z-axis, beneath the center of the instrument console. Each hatch contains a hatch-latching mechanism, which is manually operated.

A silicone elastomeric seal for each hatch is mounted in the LEM structure. When the hatch is closed, a lip near the outer circumference of the hatch enters the seal, thereby ensuring a pressure-tight contact. Because both hatches are hinged to open into the LEM, normal cabin pressurization is used to force the hatch into the seal.

After the hatches are closed and the cabin pressurized, the latch is secured manually. To open either hatch, the latch is unfastened, the cabin pressure purged through a dump valve located in the hatch proper, and the hatch is then opened.

2-5. DESCENT STAGE.

The descent stage (figure 2-4) is the unmanned portion of the LEM. If consists only of that equipment necessary for landing on the lunar surface and serves as a platform for launching the ascent stage after completion of the lunar stay. In addition to the descent engine and related components, the descent stage houses the descent control instrumentation; scientific equipment; and tanks for water, oxygen, and hydrogen used by the ECS and the EPS. The landing gear is attached externally to the descent stage.

The descent stage is constructed of aluminum alloy. The structural skin is surrounded with a layer of insulation and a very thin aluminum alloy skin that forms a modified octagonal shape around the descent stage and provides thermal protection. The outer skin is approximately 2 inches from the structural skin. Two pairs of transverse beams arranged in a cruciform, together with an upper and lower bulkhead closure, provide the main structural support. The space between the intersections of the beams forms the center compartment. Outriggers that extend from the end of each of the two pairs of beams provide attachment for the landing gear legs and support the LEM while it is housed in the spacecraft adapter. The descent stage engine is in the center compartment. Four main propellant tanks surround the engine: two oxidizer tanks between the Z-axis beams and two fuel tanks between the Y-axis beams on the lower bulkhead. Scientific equipment and helium, hydrogen, oxygen, and water tanks are adjacent to the propellant tanks.

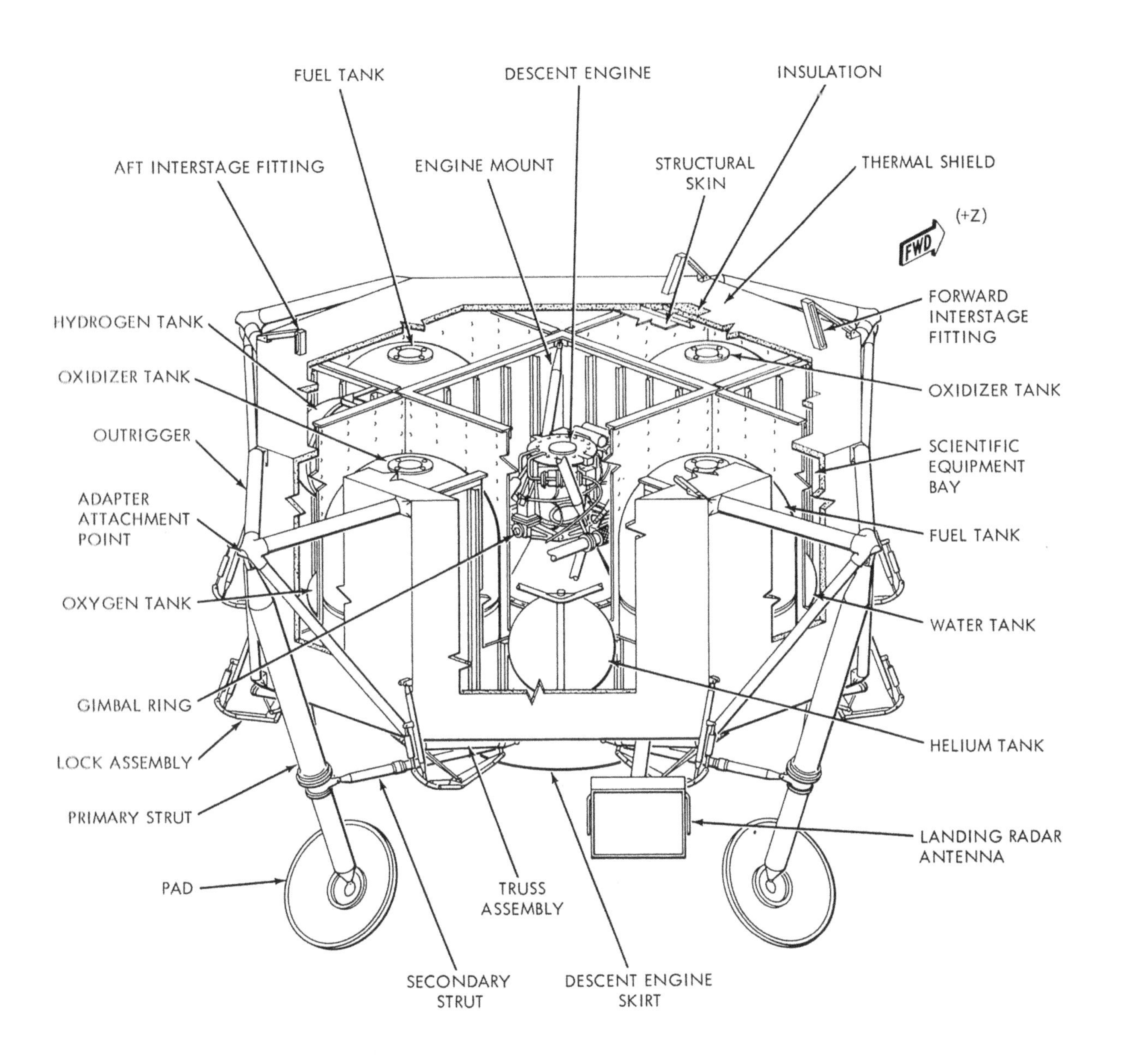

Figure 2-4. Descent Stage

The cantilever-type landing gear (Figure 2-4) consists of four legs connected to outriggers that extend from the ends of the descent stage structural beams. The legs extend from the front, rear, and sides of the LEM. Each landing gear leg consists of a primary strut and footpad, a drive-out mechanism, two secondary struts, two downlock mechanisms, and a truss. All struts have crushable attenuator inserts. The primary struts absorb compression loads; the secondary struts, compression and tension loads.

At launch, the landing gear is stowed in a retracted position and remains retracted until shortly after LEM separation from the S-IVB stage. At this time, landing gear uplocks are pyrotechnically released and springs, one contained in each drive-out mechanism, extend the landing gear. Once extended, the landing gear is locked in place by the downlock mechanism.

2-6. INTERSTAGE ATTACHMENTS, UMBILICALS, AND SEPARATIONS.

The LEM, located at launch between the Service module and the S-IVB booster, is housed within an adapter (figure 2-5) that is attached to the Service module and S-IVB booster. There is an upper and lower section to the adapter. Landing gear attachment outriggers provide for attachment of the LEM to the lower section of the adapter. Prior to transposition, the upper section of the adapter, which is in four segments, is pyrotechnically separated. These four segments, which are hinged to the lower section of the adapter, fold back. After transposition, the lower section of the adapter is pyrotechnically released, separating the adapter and S-IVB booster from the LEM.

While the LEM is docked to the Command/Service modules, umbilicals supply electrical power and pressurization to the LEM. These umbilicals are attached by the crew after transposition and are manually disconnected prior to separation of the LEM from the Command/Service modules.

Four explosive nuts and bolts connect the ascent and descent stages. At lunar launch, or for abort prior to lunar landing, the two stages are separated by the use of pyrotechnics in conjunction with ascent engine thrust. Interstage wiring umbilicals and hardlines are pyrotechnically disconnected and cryogenic piping is mechanically disconnected at stage separation.

2-7. PYROTECHNICS.

Pyrotechnic devices (figure 2-5) on board the LEM are controlled from the LEM crew compartment. The pyrotechnics that separate the adapter are controlled from the Command module. The LEM pyrotechnics are used to deploy the landing gear and antennas, to pressurize the ascent or descent Propulsion Subsystem and Reaction Control Subsystem (RCS) by opening helium valves, to transfer fuel and oxidizer between the RCS manifolds by opening crossfeed valves, and to separate the ascent and descent stages by firing the four explosive bolts and nuts that connect the two stages.

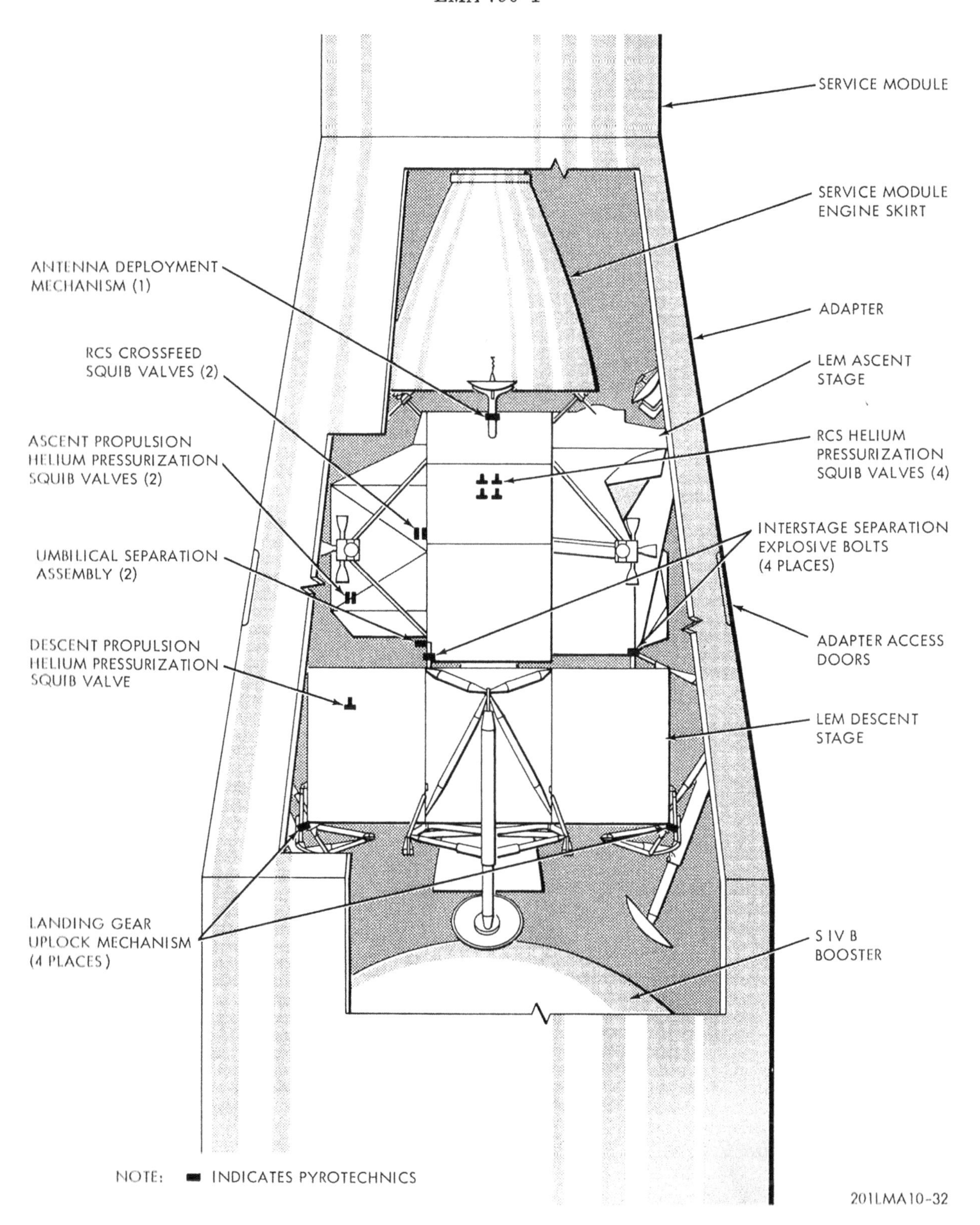

Figure 2-5. LEM Interfaces and Pyrotechnics

SECTION III

OPERATIONAL SUBSYSTEMS

3-1. GENERAL.

This section describes the LEM operational subsystems in sufficient detail to convey an understanding of the LEM as an integrated system. The integrated LEM system is comprised of the following subsystems:

Navigation and Guidance	Instrumentation
Stabilization and Control	Communications
Reaction Control	Electrical Power
Propulsion	Environmental Control

Each subsystem is functionally related to one or more of the other subsystems. A block diagram that shows the major relationships between the subsystems is presented in figure 3-1. This section also describes the LEM displays and controls that are related to all operational subsystems and crew provisions.

3-2. DISPLAYS AND CONTROLS. (See figure 3-2.)

The LEM displays are those facilities that provide information to the crewmembers. There are two main types of displays in the vehicle: visual and auditory. These displays are limited only by the fact that the crewmembers must see or hear the data provided. The LEM controls are devices that enable a change in the vehicle's status or in the crewmembers environment.

3-3. DESIGN CONSIDERATIONS.

The placement of controls and displays is such that crew safety and mission success is optimized. The majority of controls and displays that are critical, (that relate to vital parameters governing the status of the vehicle or man's safety) are centrally located, and are accessible to both crewmembers. Considerations of or trade-offs between such factors as visibility, accessibility, ease and frequency of operation or monitoring, homogeneity within panel groups, proximity to other vehicle devices or subsystems, as well as categorical division of delegated labor to each astronaut also determine individual control placement. Each crewmember is assigned specific responsibilities to assure the successful completion of all functions, tasks, or action requirements. For back-up reliability, overlapping responsibilities are assigned, to take care of a contingency situation. For example, flight displays are positioned so that they are convenient to both crew stations. Flight controls will be available to both the Commander and the Systems Engineer. This affords both astronauts the capability of controlling the spacecraft.

3-4. FLIGHT CONTROLS AND DISPLAYS.

The flight controls and displays include those components utilized in maintaining or adjusting the attitude and thrust level of the LEM during the mission, to assist in the successful landing on the lunar surface, and to provide information for rendezvous with the Command module. All primary flight displays are located on the forward

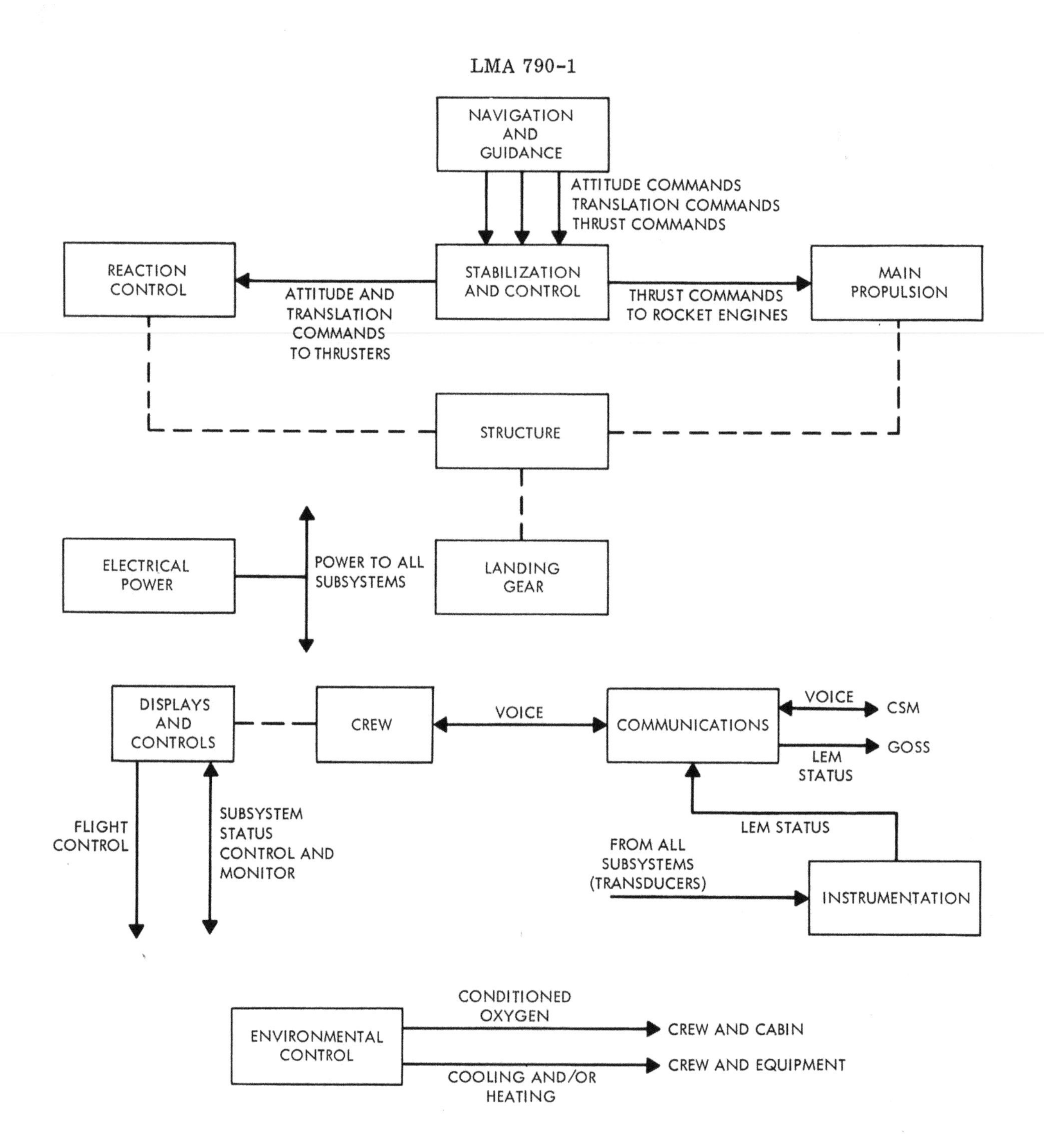

Figure 3-1. LEM Functional Block Diagram

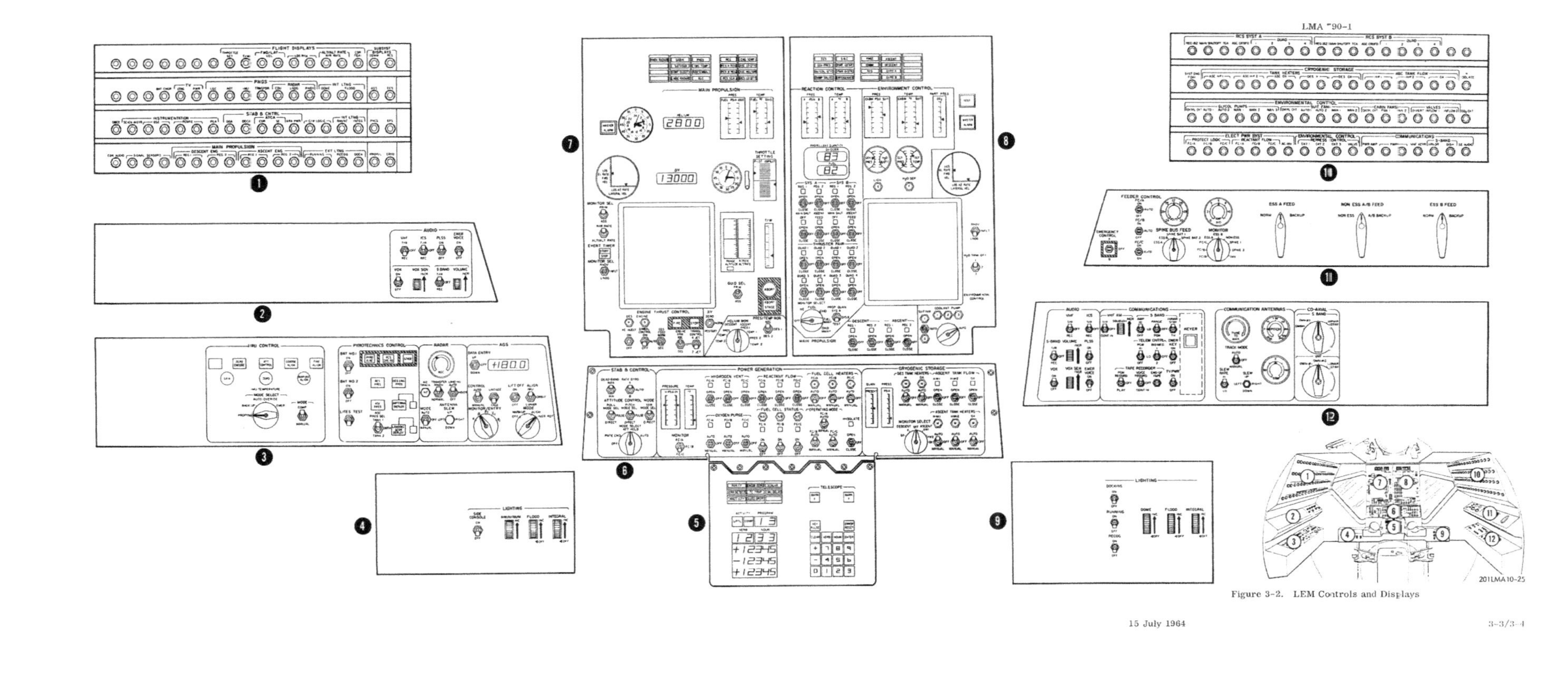

Figure 3-2. LEM Controls and Displays

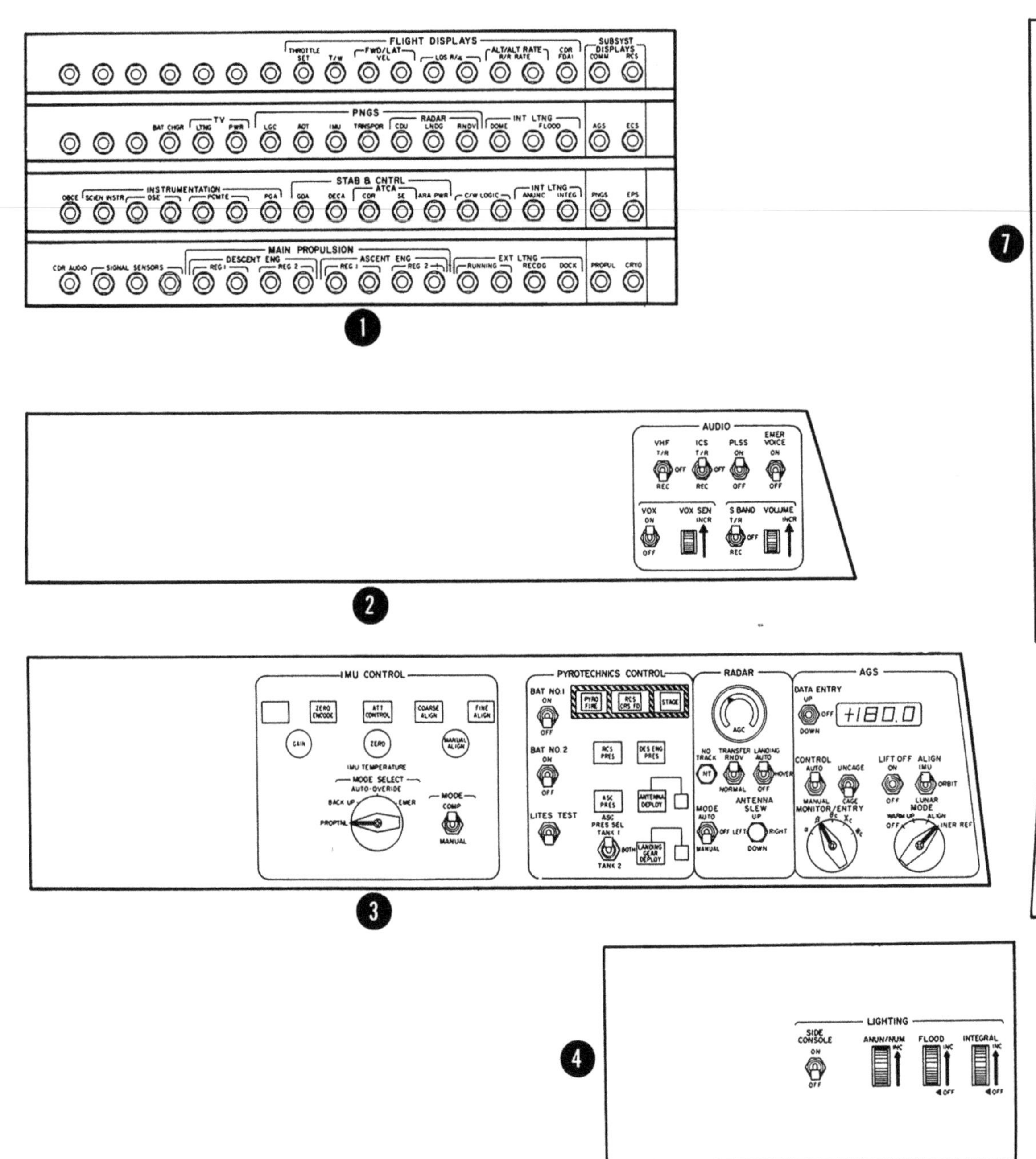

FLIGHT DISPLAYS
THROTTLE SET
T/W
FWD/LAT VEL
LOS R/A
ALT/ALT RATE R/R RATE
CDR FDAI
SUBSYST DISPLAYS
COMM
RCS
PNGS
BAT CHGR
TV
LTNG
PWR
LGC
AOT
IMU
TRANSPDR
RADAR
CDU
LNDG
RNDV
INT LTNG
DOME
FLOOD
AGS
ECS
STAB & CNTRL
INSTRUMENTATION
OBCE
SCIEN INSTR
OSE
PCMTE
PGA
GDA
DECA
ATCA
CDR
SE
ARA PWR
C/W LOGIC
ANUNC
INTEG
PNGS
EPS
MAIN PROPULSION
CDR AUDIO
SIGNAL SENSORS
DESCENT ENG
REG 1
REG 2
ASCENT ENG
EXT LTNG
RUNNING
RECOG
DOCK
PROPUL
CRYO
1
AUDIO
VHF
T/R
OFF
REC
ICS
PLSS
ON
EMER VOICE
VOX
VOX SEN
INCR
S BAND
VOLUME
2
IMU CONTROL
ZERO ENCODE
ATT CONTROL
COARSE ALIGN
FINE ALIGN
GAIN
ZERO
MANUAL ALIGN
IMU TEMPERATURE
MODE SELECT
AUTO-OVERIDE
BACK UP
EMER
PROPTNL
MODE
COMP
MANUAL
PYROTECHNICS CONTROL
BAT NO.1
BAT NO.2
PYRO FIRE
RCS CRS FD
STAGE
RCS PRES
DES ENG PRES
ASC PRES
ANTENNA DEPLOY
LITES TEST
ASC PRES SEL
TANK 1
BOTH
TANK 2
LANDING GEAR DEPLOY
RADAR
AGC
NO TRACK
TRANSFER
RNDV
LANDING
AUTO
HOVER
NORMAL
MODE
ANTENNA SLEW
UP
LEFT
RIGHT
DOWN
AGS
DATA ENTRY
DOWN
+180.0
CONTROL
UNCAGE
CAGE
LIFT OFF
ALIGN
LUNAR
ORBIT
MONITOR/ENTRY
WARM UP
INER REF
3
4
LIGHTING
SIDE CONSOLE
ANUN/NUM
INC
FLOOD
INTEGRAL

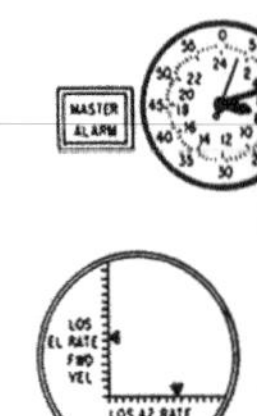

MASTER ALARM

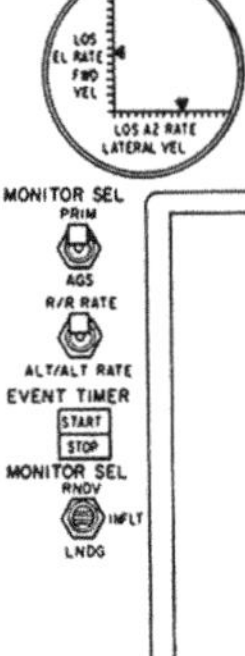

LOS EL RATE
FWD VEL
LOS AZ RATE
LATERAL VEL
MONITOR SEL
PRIM
AGS
R/R RATE
ALT/ALT RATE
EVENT TIMER
START
STOP
MONITOR SEL
RNDV
LNDG

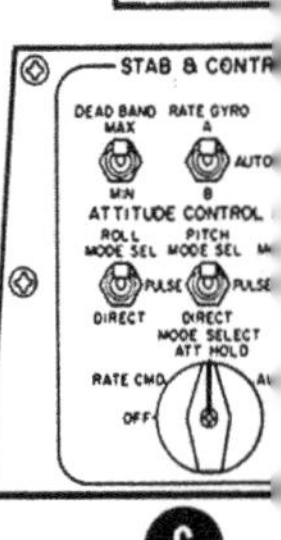

STAB & CONTR
DEAD BAND
MAX
MIN
RATE GYRO
ATTITUDE CONTROL
ROLL
MODE SEL
PITCH
PULSE
DIRECT
MODE SELECT
ATT HOLD
RATE CMD
OFF

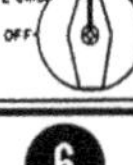

6

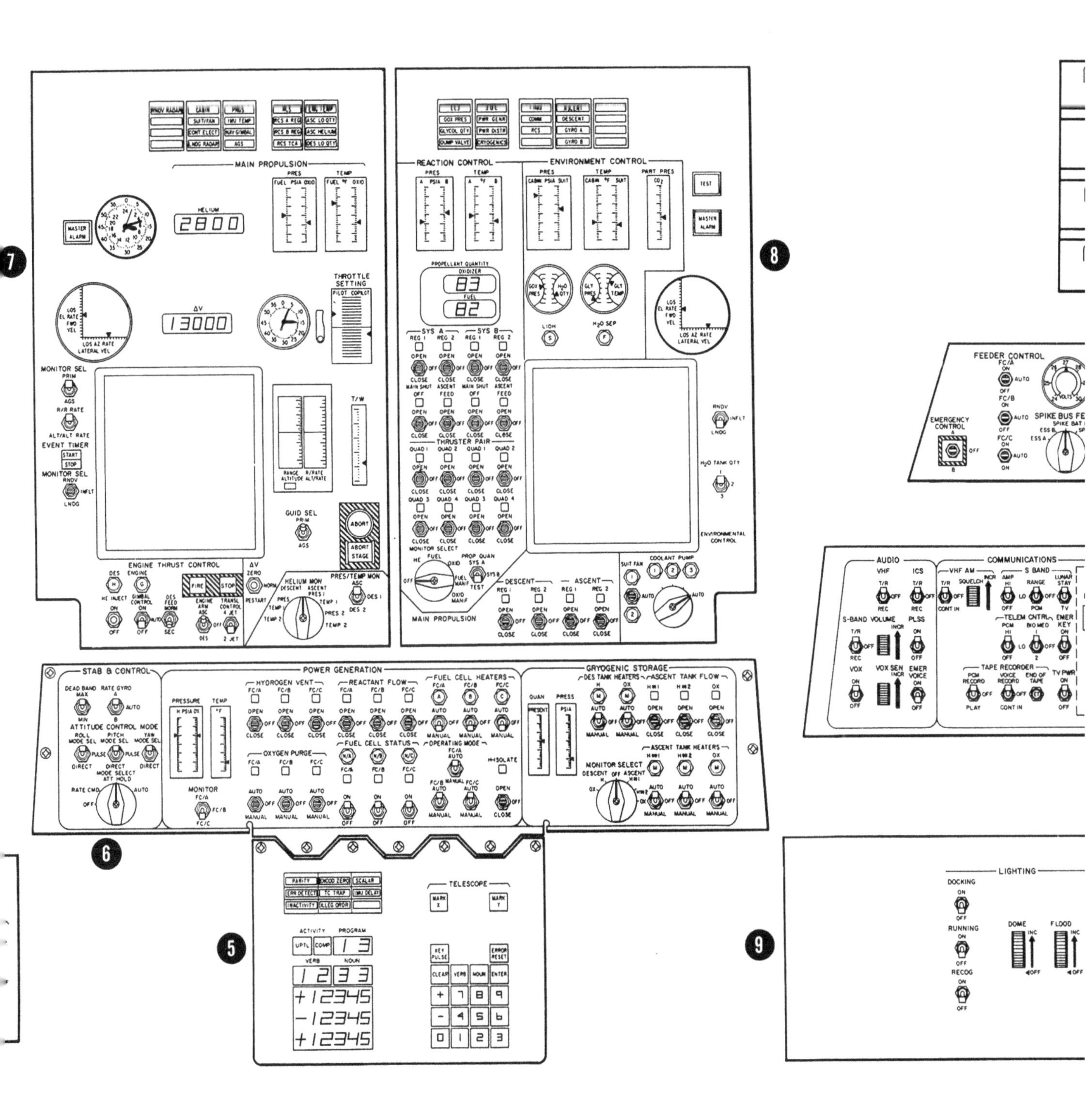

MAIN PROPULSION
REACTION CONTROL
ENVIRONMENT CONTROL
STAB & CONTROL
POWER GENERATION
CRYOGENIC STORAGE
FEEDER CONTROL
COMMUNICATIONS
TELESCOPE
LIGHTING

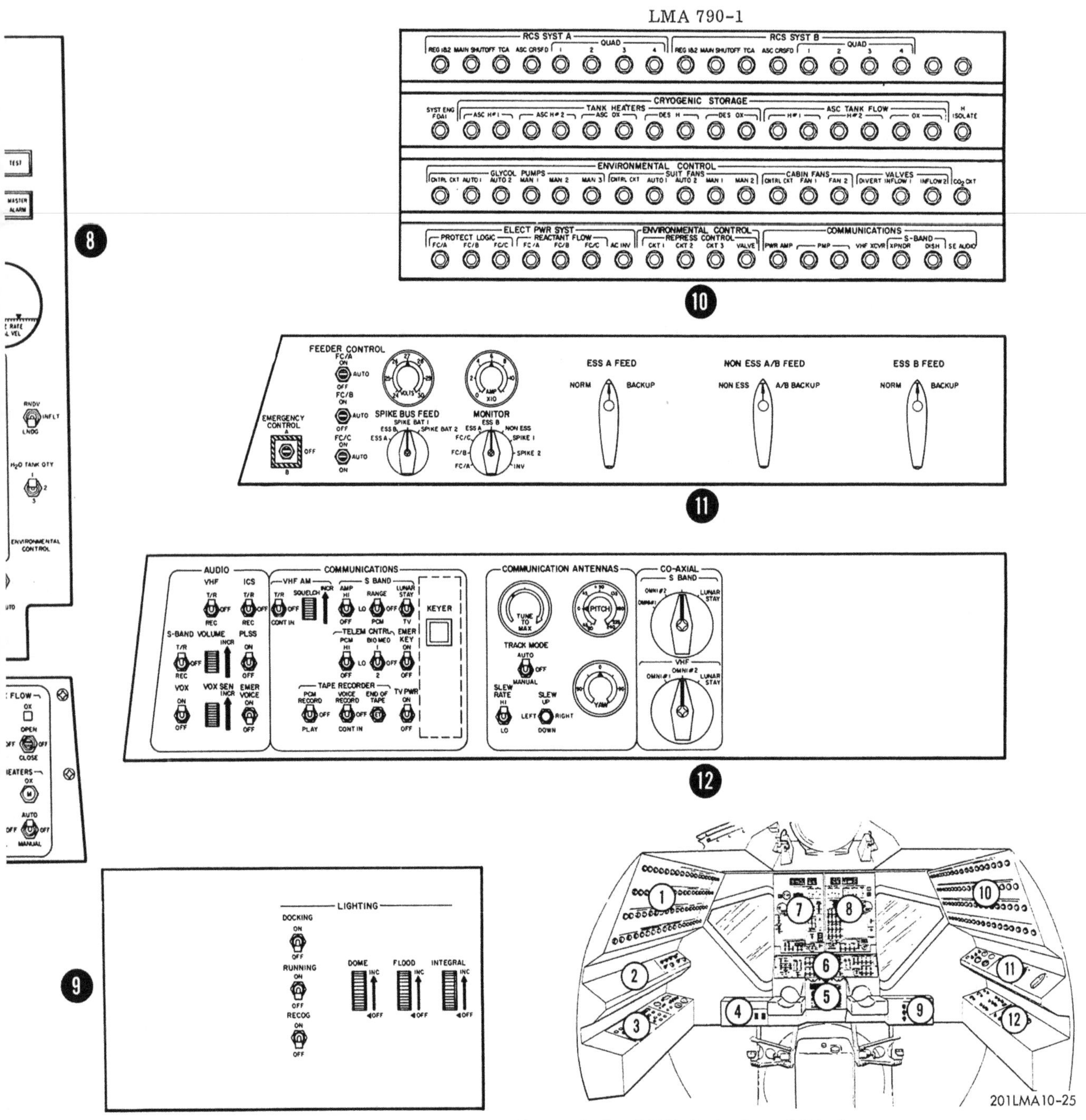

Figure 3-2. LEM Controls and Displays

instrument panel, directly in the line of sight of the Flight Commander, with repeater indicators for the Systems Engineer.

3-5. Flight Director Attitude Indicator. This indicator displays the attitudes, attitude rates, and attitude errors of the LEM, in all three axes. When used in conjunction with the appropriate control panel switch, the following information is displayed: in the landing mode of operation, the indicator displays glide slope and bearing information as well as vehicle rates and attitude; in the rendezvous mode of operation the indicator displays line of sight elevation and azimuth angles as well as vehicle rates and attitude; in the in-flight mode of operation, attitude error signals as well as rate and vehicle attitude signals are fed to the Flight Director Attitude indicator.

3-6. Forward Velocity/Lateral Velocity or LOS Azimuth Rate/LOS Elevation Rate. One of these indicators is placed on each of the crewmembers instrument panels, and provides the following information: In the landing mode of operation, the indicator displays information relative to the forward and lateral (V_z, V_y respectively) velocities of the vehicle. In the rendezvous-mode of operation, the indicator displays line of sight elevation and azimuth rates with respect to the Command module to assist in the rendezvous; in the in-flight mode of operation, no signals are fed to the indicators. A shutter apparatus is associated with the indicator to indicate the data being displayed.

3-7. Glide Slope/Bearing or LOS Azimuth/LOS Elevation. This indicator is on the Systems Engineer instrument panel; however, the indicator repeats the glide slope/bearing or LOS azimuth/LOS elevation information appearing on the Flight Director Attitude Indicator.

3-8. Thrust-To-Weight Indicator. This indicator is on the Flight Commander's instrument panel and provides the crewmembers with information relative to vehicle acceleration in the X-axis.

3-9. Δ V Accumulated. This is a digital readout instrument on the Flight Commander's instrument panel. The indicator displays changes in vehicle velocity during the various mission phases of ΔV maneuvers.

3-10. Roll Rate Indicator. This indicator is on the Flight Commander's instrument panel and displays the roll rate of the vehicle.

3-11. Pitch Rate Indicator. This instrument is on the Flight Commander's instrument panel and displays the pitch rate of the vehicle.

3-12. Yaw Rate Indicator. This instrument is on the Flight Commander's instrument panel and displays the yaw rate of the vehicle.

3-13. Range/Range Rate or Altitude/Altitude Rate Indicator. This vertical instrument is on the Flight Commanders instrument panel and works in conjunction with the Distance Mode Selector Switch. In the landing mode of operation altitude and altitude rate information is displayed to assist in the landing operation. In the rendezvous mode of operation, range and range rate with respect to the Command/Service modules is displayed. A shutter indicator is used with each of the displays to indicate the data being displayed.

3-14. Clock. A clock on the Flight Commander's panel is used to display hours, minutes and seconds on a conventional 24-hour dial face. Three pointers are incorporated in the clock face.

3-15. Events Timer. An events timer instrument on the Flight Commander's instrument panel is used to display elapsed time in minutes and seconds on a conventional dial face. The clock is actuated by a start-stop pushbutton, and can be reset by an adjacent lever arm.

3-16. LEM STATUS INDICATORS.

The operational status of many LEM subsystems are displayed on the control consoles by three types of status indicators representing three levels or degrees of importance. In decreasing order of criticality they are: (1) warning and caution lights, (2) component caution lights, and (3) status flags. These indicators provide the flight crew with supplemental information regarding the preflight readiness for the launch, the in-flight condition of the LEM, the condition of the LEM while on the lunar surface, and the LEM condition during the lunar launch.

3-17. Caution and Warning Indicators. The caution and warning equipment acts to attract the crewmembers attention to a system malfunction. The warning indicators warn of a malfunction that affects crew safety and requires an immediate action to counter the emergency. The caution indicators alert the crewmembers to a situation or malfunction that is not time critical to crew safety, but that requires that the flight crew be aware of a given problematic situation. These displays are of particular significance in subsystems which time-share displays, or in subsystems whose displays are not readily visible to the crewmembers.

In any situation where the caution or warning indicators are illuminated, the crewmembers will be able to take some constructive or corrective action to alleviate the conditions indicated. Simultaneously, when the caution or warning indicator is illuminated, information concerning the malfunction is telemetered to the ground monitoring station to ensure control station awareness of the situation in the LEM. Both master caution lights are extinguished by depressing either of the indicators. Similarly, both master warning lights are extinguished and the audible tone is silenced by depressing either of the master warning indicators. However, individual caution/warning indicators are extinguished only by a signal from the sensor at the malfunction, indicating restoration of a normal or within tolerance condition.

3-18. Component Caution Lights. The LEM advisory lights direct crew attention to abnormal conditions such as failure, manual operating mode, or other situations which result from either a malfunction or delibrate selection, and which while not imminently dangerous or critical, could jeopardize crew safety or mission success if left uncorrected or unattended.

3-19. Status Flags. Electromechanical status flags provide an immediate positive feedback regarding the physical positioning of valves or other subsystem components in response to actuation of a switch. Position information is sensed at a remote location and electrically transmitted back to position a status flag. When no response is indicated, the crewmembers are alerted to a possible failure condition. Specific status flags are provided to denote definite failure indications where applicable. A normally OPEN, ON, or DEPLOYED condition is displayed by a status flag position that is the same shade of grey as the surrounding panel. A

CLOSED, OFF, or STOWED condition is displayed by a barber pole indication; i.e., alternate black and white diagonal stripes, or, where appropriate, red, backlighted face with a black-lettered legend describing the failure.

3-20. NAVIGATION AND GUIDANCE SUBSYSTEM.

The Navigation and Guidance (N&G) Subsystem provides the measuring and data processing capabilities and control functions necessary to accomplish lunar landing and ascent, and rendezvous and docking with the Command/Service modules. The N & G Subsystem comprises the Landing Radar, the Rendezvous Radar/Transponder, the Alignment Optical Telescope, the Inertial Measurement Unit, five Coupling Data Units, the LEM Guidance Computer, and the Power and Servo Assembly. Figure 3-3 is a block diagram of the N & G Subsystem.

The N & G Subsystem is primarily an aided inertial system whose principal aids are the Landing Radar, the Rendezvous Radar/Transponder, and the Alignment Optical Telescope. The Inertial Measurement Unit is aligned to an inertial reference by star sightings with the Alignment Optical Telescope. During descent to the lunar surface, altitude and velocity information from the Landing Radar are used to update inertially derived data. During the coasting descent, lunar stay, and rendezvous phases of the mission, the Rendezvous Radar coherently tracks its transponder on the Command module to provide range, range rate, and angle rate measurements (with respect to inertial space) to the LEM Guidance Computer.

3-21. LANDING RADAR.

The Landing Radar (LR) senses LEM velocity and altitude with respect to the lunar surface, both when the spacecraft is moving tangentially to the lunar surface, and when it rotates to a vertical attitude to complete its final descent. Velocity and altitude information are applied to the LEM Guidance Computer, where they are used to check or update inertially derived data, and are displayed during descent from an altitude of 20,000 feet to touchdown. The LR comprises an antenna assembly, an electronics assembly, and a control assembly, and is functionally divided into a three-beam, continuous wave (CW) doppler velocity sensor and a narrow beam, linear FM/CW radar altimeter. Figure 3-4 is a block diagram of the Landing Radar.

The antenna assembly comprises a space-duplexed array of transmit and receive antennas on which are mounted the solid state transmitters, modulator, detectors, preamplifiers, test modulators, and waveguides. The transmit array generates four beams. Three of these are arranged in a lambda configuration and are used by the doppler velocity sensor; the fourth is used by the radar altimeter. The receiving antennas consist of four individual broad-side arrays. The receiving array beamwidths are wider than those of the transmit array so that antenna boresighting is not critical. The electronics assembly contains the frequency trackers, frequency-to-analog converters, and the power supply. It provides outputs to the LEM Guidance Computer in the form of pulse trains that correspond to the doppler frequencies (D_1, D_2, and D_3) and the range (along the altitude beam) frequency (f_r). Analog outputs to the control assembly permit display of the spacecraft velocity components (V_x, V_y, and V_z), and range along the altitude beam (R). The LR supplies accurate data from 25,000 feet to touchdown without mode changes or altitude holds, and has provision for hovering and negative speeds. The LR self-test devices enable operational checks of the entire LR (including the radar front end) without radar return from the lunar surface, so that the astronauts may evaluate the operational status of the LR at any time while in earth orbit, translunar flight, or lunar orbit.

3-22. Doppler Velocity Sensor. The doppler velocity sensor is characterized by its three narrow beams of CW radiation, a solid-state transmitter, direct-to-audio detection, frequency trackers, and frequency-to-dc converters to provide the desired doppler frequencies and spacecraft velocity component outputs.

The received energy from each beam is detected in phase quadrature in dual microwave detectors using the direct-to-audio detection technique. The local oscillator signal is supplied by the solid-state transmitter; no separate local oscillator or i-f stages are necessary. The received signals, detected in quadrature to retain sign sense, are amplified in two-stage, gain-switching preamplifiers. Unwanted transmitter leakage is heterodyned to zero and inherently rejected, as the detectors are a-c coupled to the preamplifiers. The amplified quadrature doppler signals for each beam are then applied to velocity sensor frequency trackers, which search the band of expected doppler frequencies with a narrow beamwidth filter. An acquisition circuit is energized when the doppler signal appears in the tracker band, and the tracker locks on and continuously tracks and filters the doppler spectrum. The output of the three velocity sensor frequency trackers, which are 600 kc carriers with positive doppler signals riding above and negative doppler signals below ($f_c + D_1$, $f_c + D_2$, and $f_c + D_3$), are applied in pulse train form to the LEM Guidance Computer, and to the arithmetic unit. They are combined in the arithmetic unit to provide output frequencies proportional to the velocity components along the three orthogonal coordinates of the antenna system (V_{xa}, V_{ya}, and V_{za}), and applied to the frequency converters. Here, they are converted to d-c voltages in a pulse train format. The polarity of the pulse train denotes the sign of the represented antenna velocity component. In the coordinate converter, the spacecraft velocity components (V_x, V_y, and V_z) are derived from the antenna velocity components and the direction cosines of the antenna pointing angles. These are applied to the control assembly for display.

3-23. Radar Altimeter. The radar altimeter employs narrow beam, linear FM/CW radiation, a solid-state transmitter, a frequency tracker, and a converter to provide outputs of range along the altitude beam.

The received energy is detected in phase quadrature with balanced mixers using the direct-to-audio detection method. In a manner almost identical to that of the velocity sensor, the quadrature signals are amplified in a two-stage, gain-switching preamplifier, and applied to the altimeter frequency tracker. The frequency along the range beam is the sum of the range frequency and the doppler frequency ($f_r + f_d$). The doppler component is removed in the arithmetic unit by mixing operations, and the range frequency signal is applied to the altitude frequency converter, where the range signals are derived.

3-24. RENDEZVOUS RADAR/TRANSPONDER.

The Rendezvous Radar (RR) is an X-band, interrupted-constant-wave, amplitude comparison, monopulse radar that can acquire and accurately track its Transponder in the Command module at any range between 400 nautical miles and 5 feet. (An

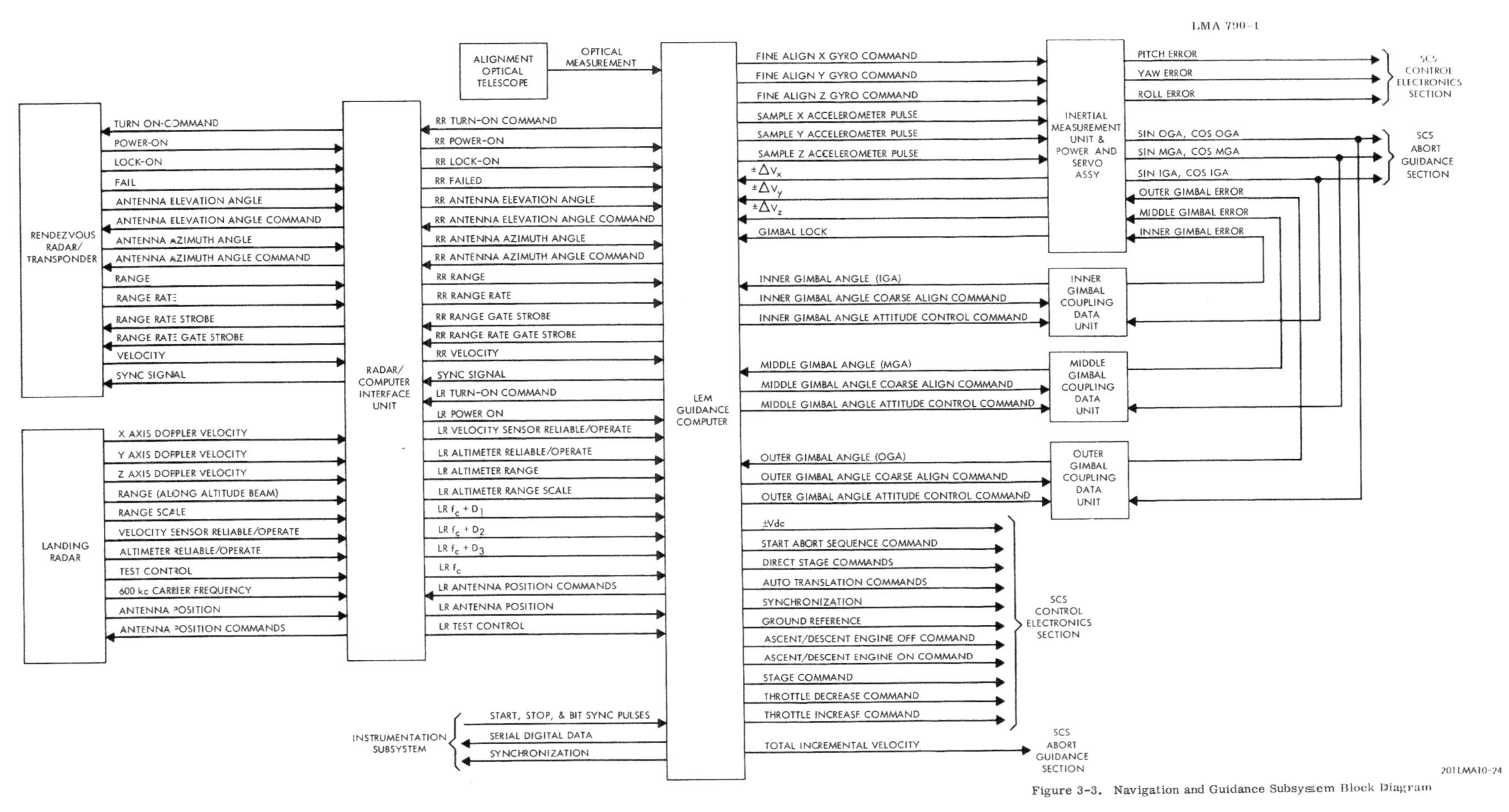

Figure 3-3. Navigation and Guidance Subsystem Block Diagram

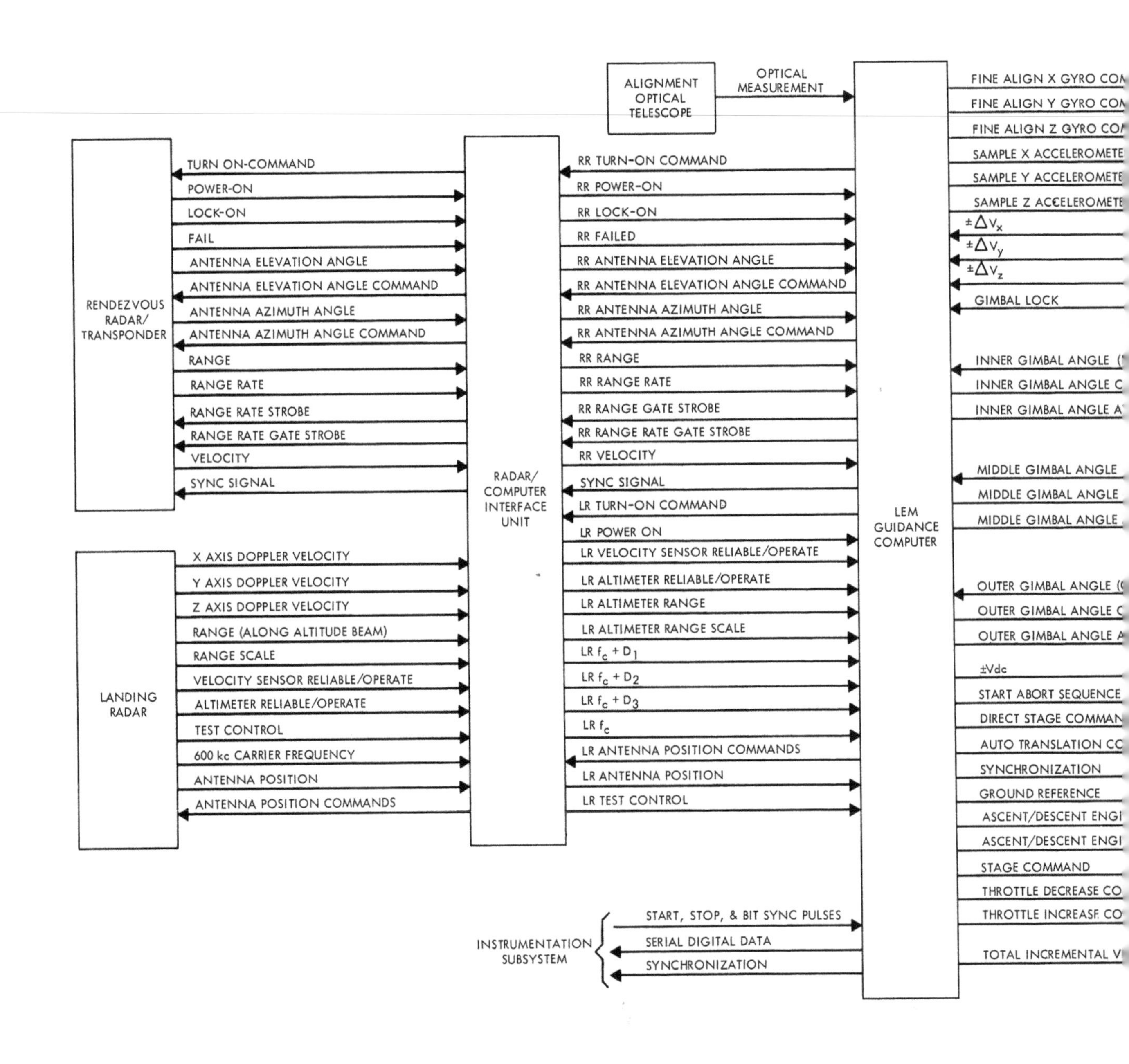

ALIGNMENT OPTICAL TELESCOPE
OPTICAL MEASUREMENT
RENDEZVOUS RADAR/ TRANSPONDER
TURN ON-COMMAND
POWER-ON
LOCK-ON
FAIL
ANTENNA ELEVATION ANGLE
ANTENNA ELEVATION ANGLE COMMAND
ANTENNA AZIMUTH ANGLE
ANTENNA AZIMUTH ANGLE COMMAND
RANGE
RANGE RATE
RANGE RATE STROBE
RANGE RATE GATE STROBE
VELOCITY
SYNC SIGNAL
LANDING RADAR
X AXIS DOPPLER VELOCITY
Y AXIS DOPPLER VELOCITY
Z AXIS DOPPLER VELOCITY
RANGE (ALONG ALTITUDE BEAM)
RANGE SCALE
VELOCITY SENSOR RELIABLE/OPERATE
ALTIMETER RELIABLE/OPERATE
TEST CONTROL
600 kc CARRIER FREQUENCY
ANTENNA POSITION
ANTENNA POSITION COMMANDS
RADAR/ COMPUTER INTERFACE UNIT
RR TURN-ON COMMAND
RR POWER-ON
RR LOCK-ON
RR FAILED
RR ANTENNA ELEVATION ANGLE
RR ANTENNA ELEVATION ANGLE COMMAND
RR ANTENNA AZIMUTH ANGLE
RR ANTENNA AZIMUTH ANGLE COMMAND
RR RANGE
RR RANGE RATE
RR RANGE GATE STROBE
RR RANGE RATE GATE STROBE
RR VELOCITY
SYNC SIGNAL
LR TURN-ON COMMAND
LR POWER ON
LR VELOCITY SENSOR RELIABLE/OPERATE
LR ALTIMETER RELIABLE/OPERATE
LR ALTIMETER RANGE
LR ALTIMETER RANGE SCALE
LR $f_c + D_1$
LR $f_c + D_2$
LR $f_c + D_3$
LR f_c
LR ANTENNA POSITION COMMANDS
LR ANTENNA POSITION
LR TEST CONTROL
LEM GUIDANCE COMPUTER
INSTRUMENTATION SUBSYSTEM
START, STOP, & BIT SYNC PULSES
SERIAL DIGITAL DATA
SYNCHRONIZATION
FINE ALIGN X GYRO CO
FINE ALIGN Y GYRO CO
FINE ALIGN Z GYRO CO
SAMPLE X ACCELEROMETE
SAMPLE Y ACCELEROMETE
SAMPLE Z ACCELEROMETE
$\pm\Delta V_x$
$\pm\Delta V_y$
$\pm\Delta V_z$
GIMBAL LOCK
INNER GIMBAL ANGLE
INNER GIMBAL ANGLE C
INNER GIMBAL ANGLE A
MIDDLE GIMBAL ANGLE
MIDDLE GIMBAL ANGLE
MIDDLE GIMBAL ANGLE
OUTER GIMBAL ANGLE
OUTER GIMBAL ANGLE C
OUTER GIMBAL ANGLE A
±Vdc
START ABORT SEQUENCE
DIRECT STAGE COMMAN
AUTO TRANSLATION CO
SYNCHRONIZATION
GROUND REFERENCE
ASCENT/DESCENT ENGI
ASCENT/DESCENT ENGI
STAGE COMMAND
THROTTLE DECREASE CO
THROTTLE INCREASE CO
TOTAL INCREMENTAL V

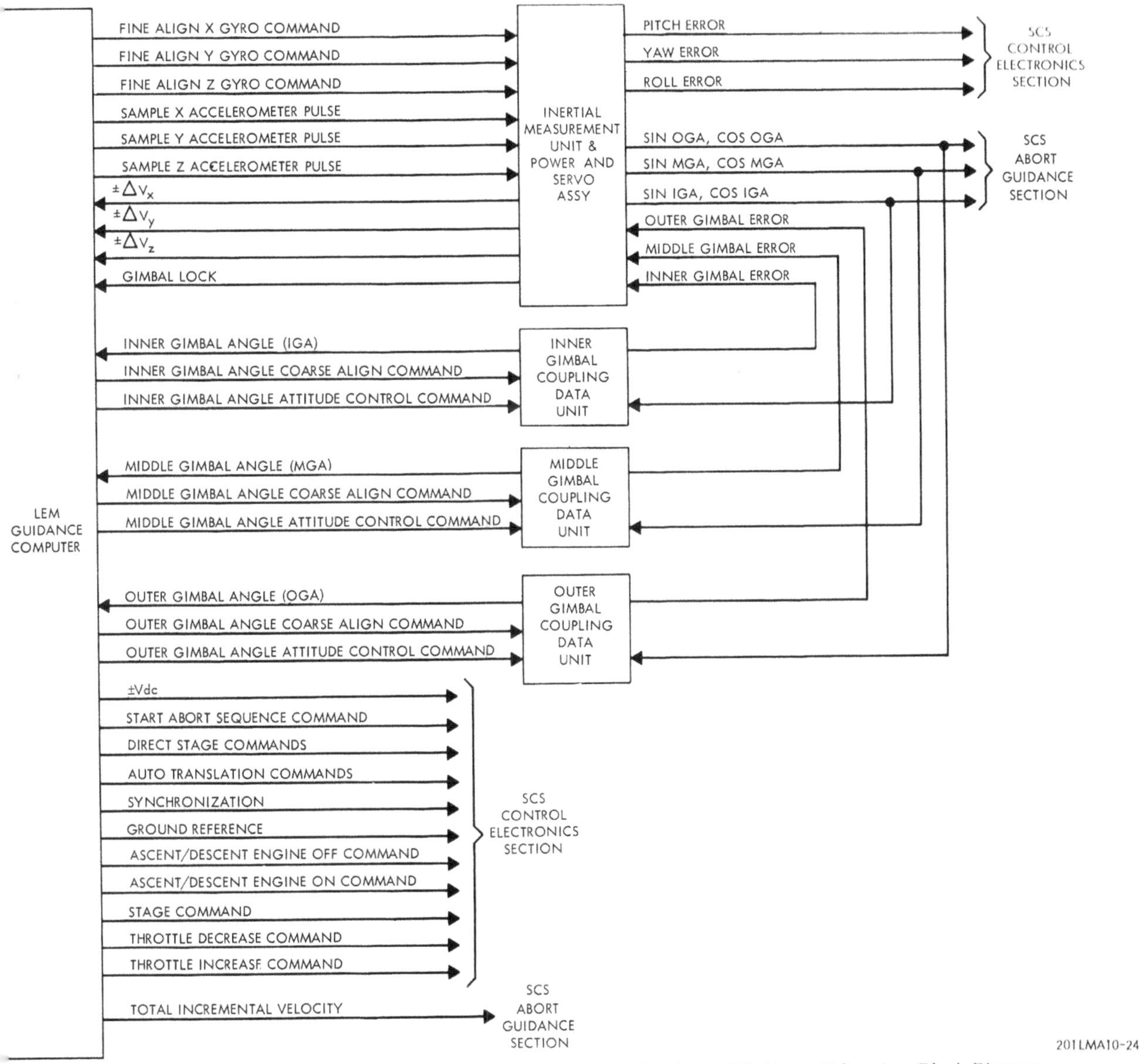

LEM GUIDANCE COMPUTER

FINE ALIGN X GYRO COMMAND
FINE ALIGN Y GYRO COMMAND
FINE ALIGN Z GYRO COMMAND
SAMPLE X ACCELEROMETER PULSE
SAMPLE Y ACCELEROMETER PULSE
SAMPLE Z ACCELEROMETER PULSE
$\pm\Delta V_x$
$\pm\Delta V_y$
$\pm\Delta V_z$
GIMBAL LOCK

INERTIAL MEASUREMENT UNIT & POWER AND SERVO ASSY

PITCH ERROR
YAW ERROR
ROLL ERROR

SCS CONTROL ELECTRONICS SECTION

SIN OGA, COS OGA
SIN MGA, COS MGA
SIN IGA, COS IGA

SCS ABORT GUIDANCE SECTION

OUTER GIMBAL ERROR
MIDDLE GIMBAL ERROR
INNER GIMBAL ERROR

INNER GIMBAL ANGLE (IGA)
INNER GIMBAL ANGLE COARSE ALIGN COMMAND
INNER GIMBAL ANGLE ATTITUDE CONTROL COMMAND

INNER GIMBAL COUPLING DATA UNIT

MIDDLE GIMBAL ANGLE (MGA)
MIDDLE GIMBAL ANGLE COARSE ALIGN COMMAND
MIDDLE GIMBAL ANGLE ATTITUDE CONTROL COMMAND

MIDDLE GIMBAL COUPLING DATA UNIT

OUTER GIMBAL ANGLE (OGA)
OUTER GIMBAL ANGLE COARSE ALIGN COMMAND
OUTER GIMBAL ANGLE ATTITUDE CONTROL COMMAND

OUTER GIMBAL COUPLING DATA UNIT

±Vdc
START ABORT SEQUENCE COMMAND
DIRECT STAGE COMMANDS
AUTO TRANSLATION COMMANDS
SYNCHRONIZATION
GROUND REFERENCE
ASCENT/DESCENT ENGINE OFF COMMAND
ASCENT/DESCENT ENGINE ON COMMAND
STAGE COMMAND
THROTTLE DECREASE COMMAND
THROTTLE INCREASE COMMAND

SCS CONTROL ELECTRONICS SECTION

TOTAL INCREMENTAL VELOCITY

SCS ABORT GUIDANCE SECTION

201LMA10-24

Figure 3-3. Navigation and Guidance Subsystem Block Diagram

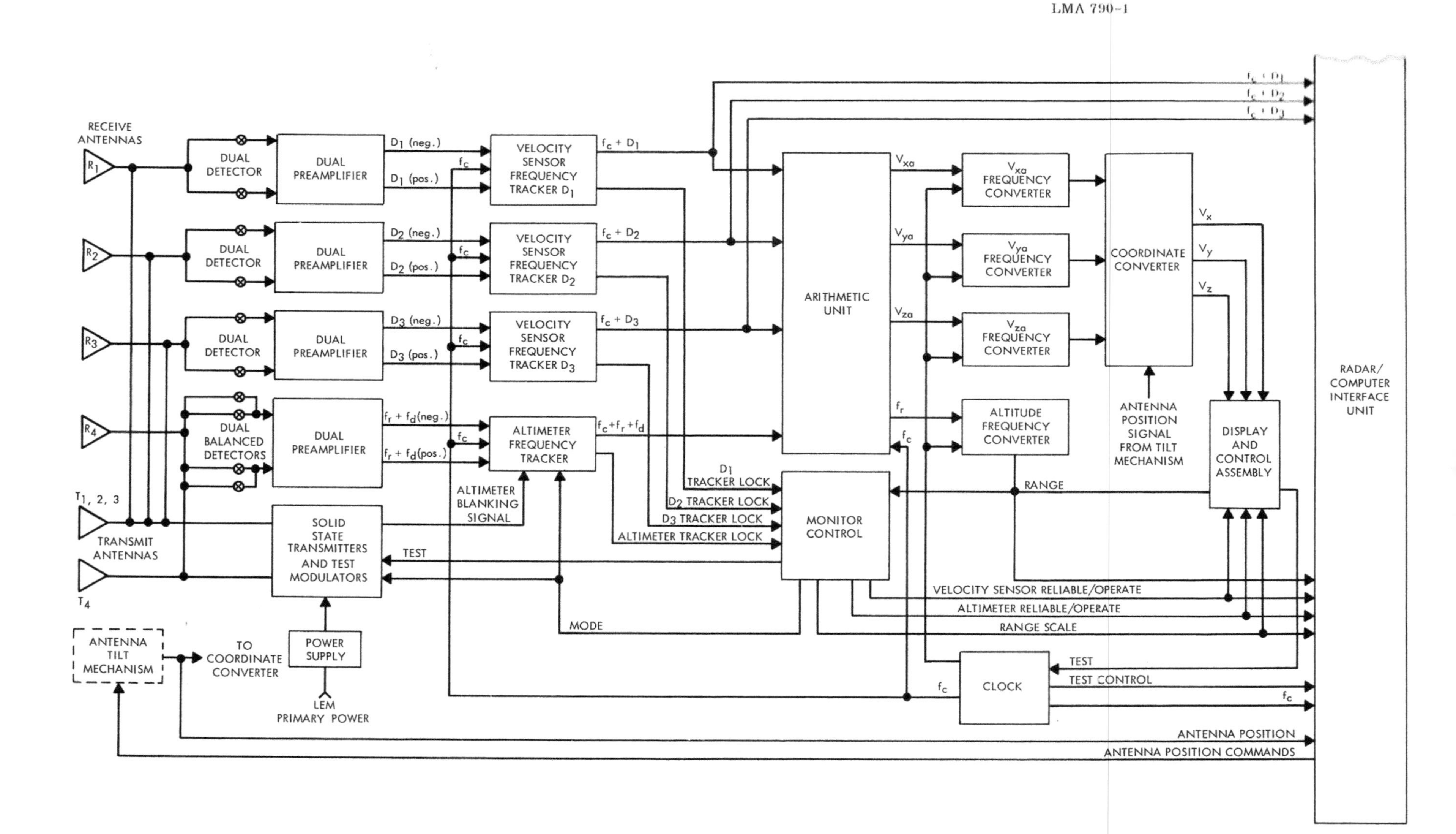

Figure 3-4. Landing Radar Block Diagram

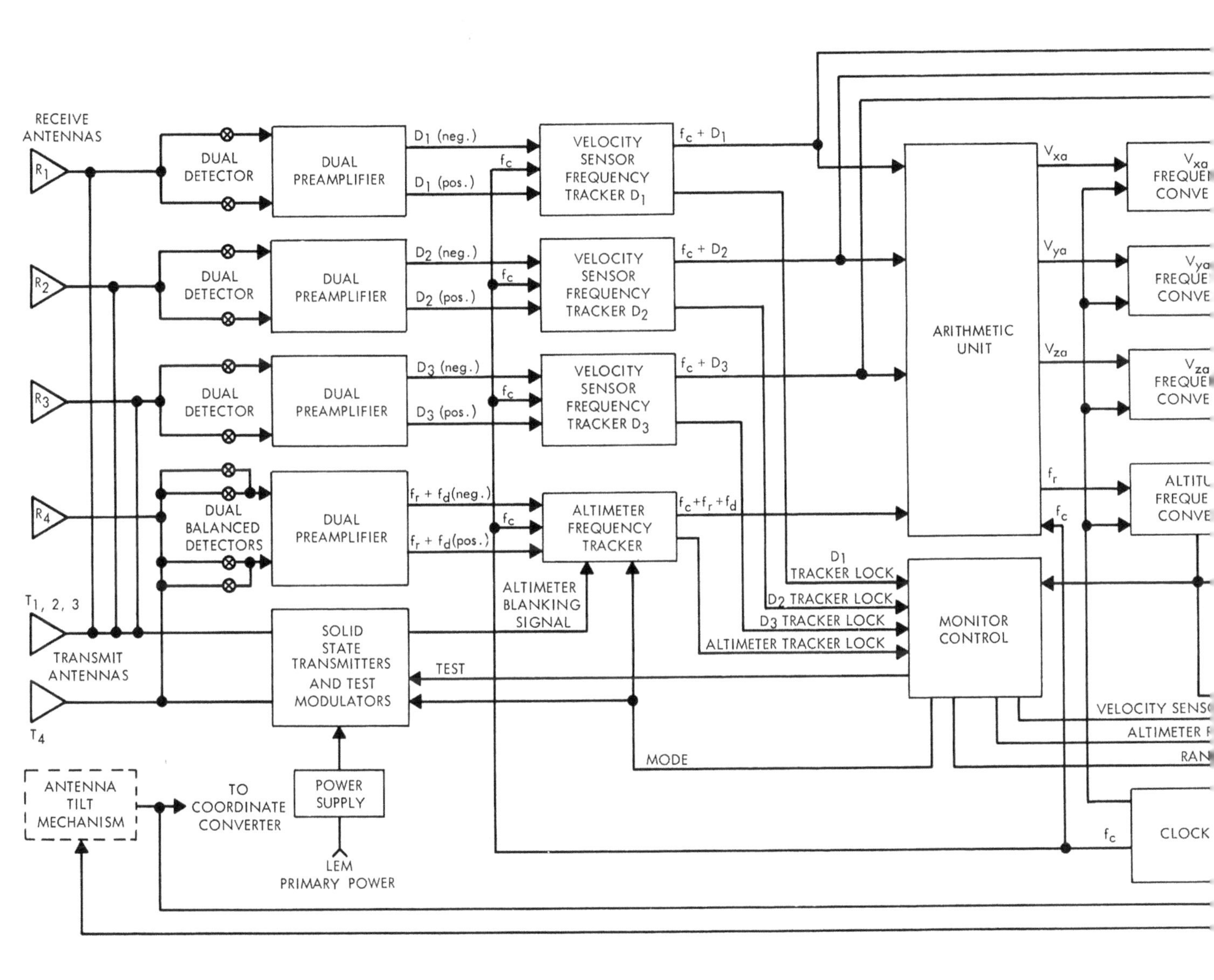

RECEIVE ANTENNAS
R1
R2
R3
R4
DUAL DETECTOR
DUAL DETECTOR
DUAL DETECTOR
DUAL BALANCED DETECTORS
DUAL PREAMPLIFIER
DUAL PREAMPLIFIER
DUAL PREAMPLIFIER
DUAL PREAMPLIFIER
D1 (neg.)
D1 (pos.)
D2 (neg.)
D2 (pos.)
D3 (neg.)
D3 (pos.)
fr + fd(neg.)
fr + fd(pos.)
fc
VELOCITY SENSOR FREQUENCY TRACKER D1
VELOCITY SENSOR FREQUENCY TRACKER D2
VELOCITY SENSOR FREQUENCY TRACKER D3
ALTIMETER FREQUENCY TRACKER
fc + D1
fc + D2
fc + D3
fc+fr+fd
ARITHMETIC UNIT
Vxa
Vya
Vza
fr
fc
D1 TRACKER LOCK
D2 TRACKER LOCK
D3 TRACKER LOCK
ALTIMETER TRACKER LOCK
MONITOR CONTROL
ALTIMETER BLANKING SIGNAL
T1, 2, 3
TRANSMIT ANTENNAS
T4
SOLID STATE TRANSMITTERS AND TEST MODULATORS
TEST
MODE
POWER SUPPLY
LEM PRIMARY POWER
ANTENNA TILT MECHANISM
TO COORDINATE CONVERTER
CLOCK
VELOCITY SENS
ALTIMETER
RAN

Figure 3-4.

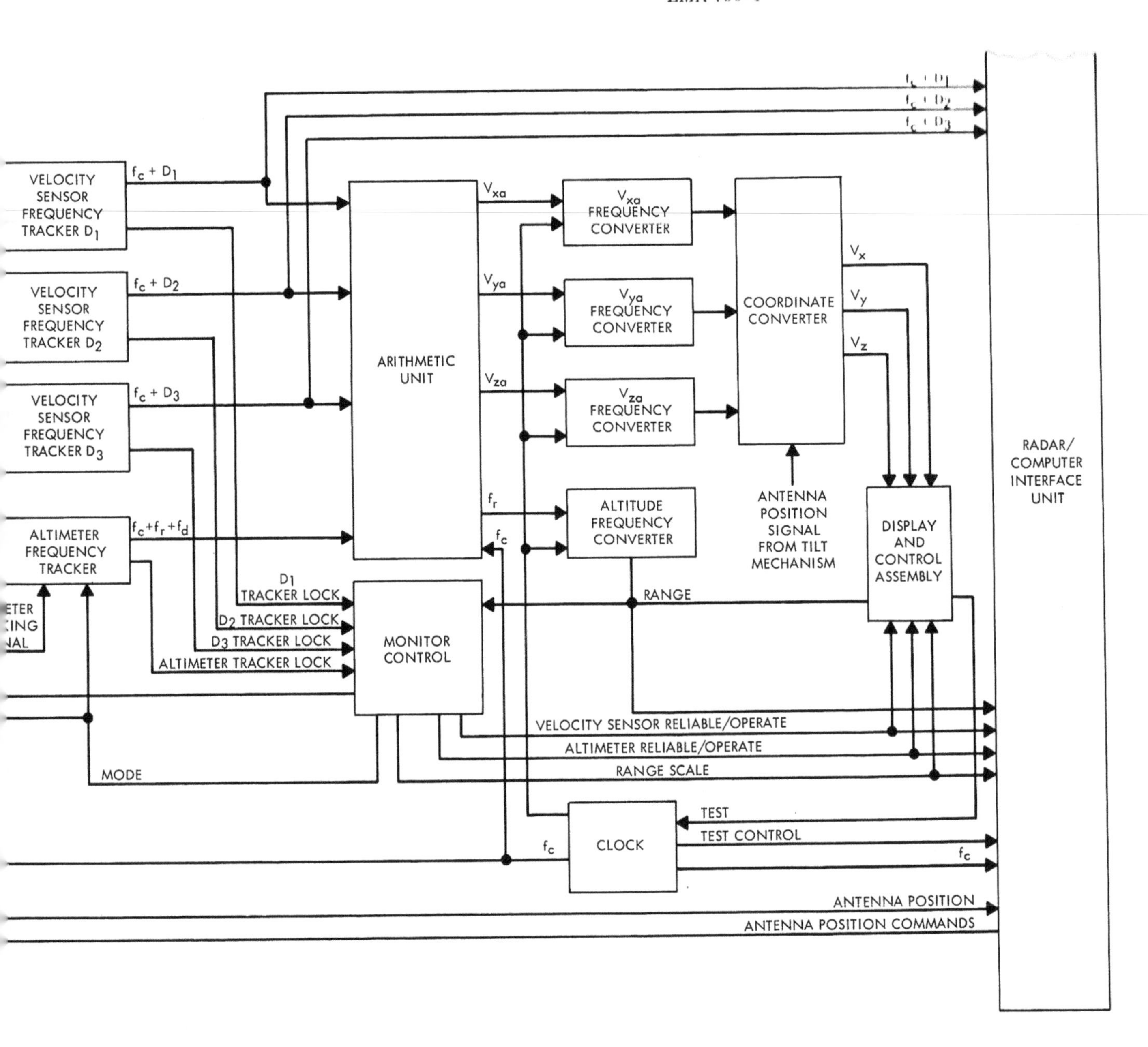

Figure 3-4. Landing Radar Block Diagram

identical Transponder in the LEM is tracked by the Command module.) Range rate measurements are provided at any LEM rate of closure or opening (with respect to the Command Module) between 4,900 feet per second and 1 foot per second. Figure 3-5 is a simplified block diagram of the Rendezvous Radar.

The Rendezvous Radar antenna assembly includes a gyro-stabilized monopulse antenna; sum-and-difference hybrid networks; a duplexer; a high-level, solid-state varactor-multiplier-transmitter; a low-power, solid-state, varactor-multiplier-local oscillator; three mixers; and three i-f preamplifiers. The elevation error receiver, sum channel receiver, and azimuth error receiver are identical i-f assemblies that convert the one reference and two error signals to an i-f of 1,000 kc. Within each i-f assembly is a 1-kc filter that provides approximately 13 db signal-to-noise ratio improvement for angle tracking. The sum channel receiver output is used as a reference signal for the two phase-sensitive detectors within the angle track module. These detectors convert the output of the elevation and azimuth error receivers to two bipolar-video angle error signals and apply them to boxcar detectors, which provide the ΔEL and ΔAZ bipolar d-c error signals for antenna line-of-sight control.

The range tracker has two modes of operation: the variable-PRF, closed-loop mode is used for ranges between 1,000 feet and 400 miles, and the fixed-PRF, open-loop mode is used for ranges less than 1,000 feet. In either mode, Transponder returns or surface returns can be used to derive range. In the variable-PRF mode, the range derived by the range tracker is used to control the PRF; in the fixed PRF mode, the PRF is locked at 250 kc. The frequency tracker measures the doppler-shifted response of Transponder or surface returns to derive velocity. The frequency synthesizer generates the various frequencies required by the transmitter and local oscillator chains in the various modes of operation. The LEM Guidance Computer antenna scan commands, tracking error signals, and gyro signals are combined by the antenna servo to produce antenna drive signals. Signals proportional to antenna line-of-sight angles are applied to the radar/computer interface units for transfer to the LEM Guidance Computer.

The radar/computer interface units provide the interfaces between the Landing and Rendezvous Radars and the LEM Guidance Computer. Radar output signals are processed into the format required by the LEM Guidance Computer and selected sequentially for transfer by coded strobe signals generated by the computer.

3-25. ALIGNMENT OPTICAL TELESCOPE.

The Alignment Optical Telescope (AOT) is a non-articulating, unit power, 60-degree field-of-view telescope used to determine the position and orientation of the spacecraft for aligning the Inertial Measurement Unit. The AOT provides the astronauts with data obtained by measuring angles between lines of sight to celestial bodies to establish the inertial reference. The telescope optics mechanism has three distinct viewing positions and a stowage position. Data obtained from the use of the AOT is manually entered into the LEM Guidance Computer.

3-26. INERTIAL MEASUREMENT UNIT.

The Inertial Measurement Unit (IMU) is the primary inertial sensing device of the LEM. Three integrating gyroscopes and three pendulous accelerometers are mounted on the innermost gimbal of a three-degree-of-freedom gimbal system.

The innermost gimbal is held non-rotating with respect to inertial space by the three gimbal servos, which derive their input error signals from the three gyroscopes. The orthogonal input axes of the three accelerometers represent the LEM X, Y, and Z axes. Velocity changes along these axes are sensed by the accelerometers, which produce signals representing incremental changes in velocity and apply them to the LEM Guidance Computer for the calculation of total velocity. Signals proportional to the attitude of the LEM are generated by the inner, middle, and outer gimbal angle resolvers. These signals are applied to the LEM Guidance Computer and to the astronaut displays. The LEM Guidance Computer calculates gimbal angles that would result from bringing the LEM to a desired attitude, and compares these with the actual gimbal angles supplied by the gimbal angle resolvers. Any difference between the actual and calculated gimbal angles results in the generation of steering error signals, which are applied to the Stabilization and Control Subsystem to correct the attitude of the LEM.

3-27. COUPLING DATA UNIT.

The Coupling Data Units (CDU's) convert and transfer angular information between major elements of the N & G Subsystem and the Stabilization and Control Subsystem (SCS). The N & G Subsystem uses five CDU's, one for the shaft axis and one for the trunnion axis of the Rendezvous Radar antenna, and one for each of the three gimbals of the IMU.

The two CDU's used with the Rendezvous Radar antenna provide interfaces between the antenna and the LEM Guidance Computer. The LEM Guidance Computer calculates digital antenna position commands prior to acquisition of the Command Module Transponder. These signals are converted into analog form by the CDU's, and applied to the antenna drive mechanism, causing the antenna to be aimed at the orbiting Command/Service modules. Following Transponder acquisition, Command/Service module tracking information is digitized by the CDU's and applied to the LEM Guidance Computer.

The three CDU's working with the IMU provide interfaces between the IMU and the LEM Guidance Computer, and between the LEM Guidance Computer and the SCS. Each IMU gimbal angle resolver provides its CDU with analog gimbal angle signals representing the attitude of the LEM. The CDU coverts these signals to digital form and applies them to the LEM Guidance Computer. Using these signals, the computer will calculate any necessary steering error signals and route them to the proper CDU in digital form. The CDU then converts these to 800 cps analog signals and applies them to the SCS. IMU coarse and fine align commands, which are generated by the LEM Guidance Computer, are coupled to the IMU by the CDU's.

The digital CDU principle is based upon the use of a-c analog computing techniques to derive error voltages from analog resolver angle signals. Digital circuitry continuously alters the analog computational scaling factor to null the error voltages. When this null occurs, the state of the digital memory drive circuit exactly corresponds to the resolver position.

3-28. LEM GUIDANCE COMPUTER.

The LEM Guidance Computer (LGC) is the central data processing device of the N & G Subsystem. It is a parallel, fixed point, one's complement, general-purpose digital computer with a fixed rope core memory, an erasable ferrite core memory, and a

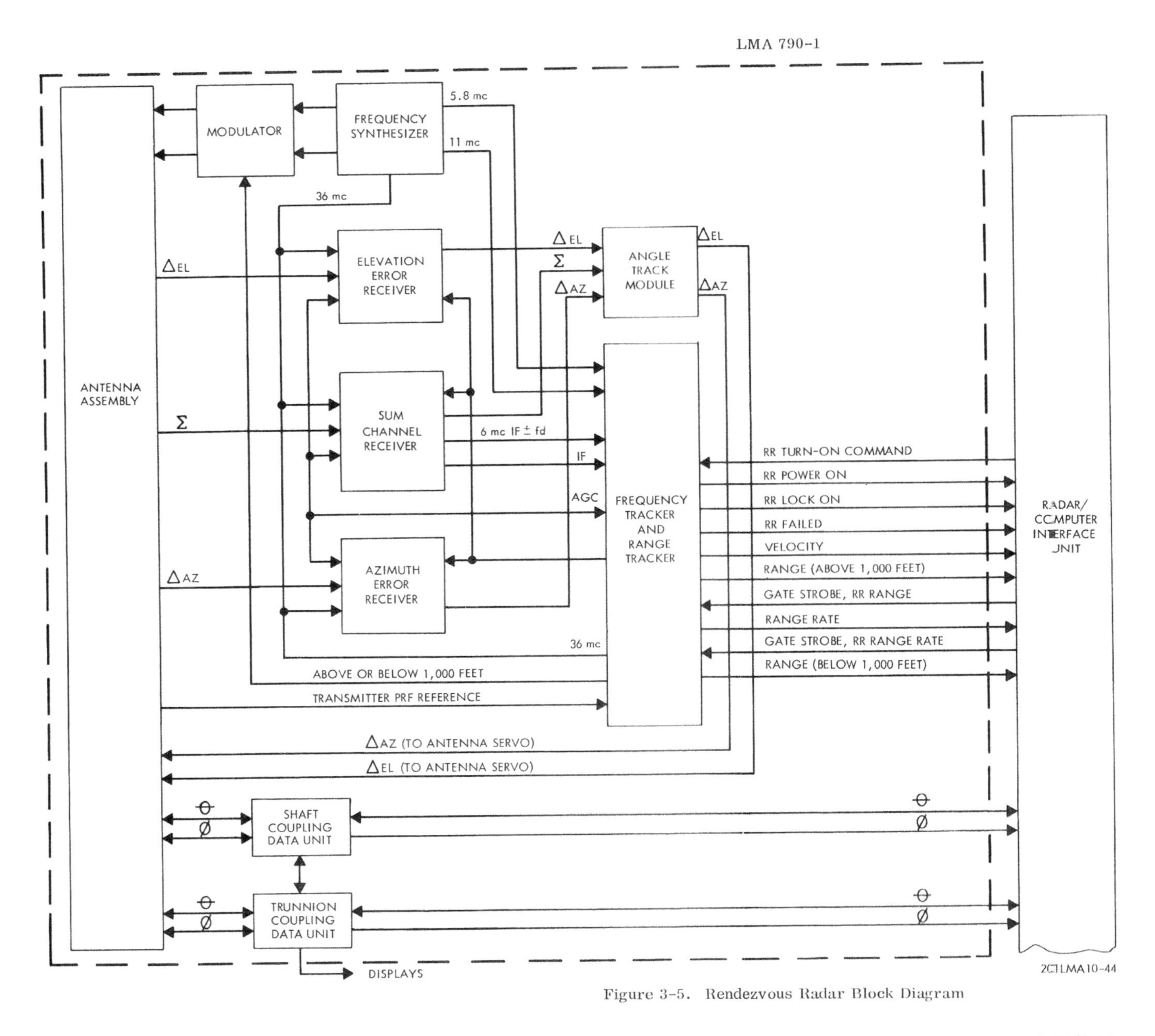

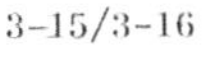

Figure 3-5. Rendezvous Radar Block Diagram

limited self-check capability. Inputs to the LGC are received from the Landing and Rendezvous Radars through the radar/computer interface units, from the AOT through the control panel, from the IMU through the CDU's, and from the navigator via the data entry keyboard on the LGC control panel. The LGC performs four major functions: (1) calculate steering signals and generate engine discretes necessary to keep the spacecraft on a required trajectory, (2) align the stable member (innermost gimbal) of the IMU to a coordinate system defined by precise optical measurements, (3) conduct limited malfunction isolation of the N & G Subsystem, and (4) display pertinent navigation information to the astronauts. Using information from navigation fixes, the LGC determines the amount of deviation from the required trajectory and calculates the necessary attitude and thrust corrective commands. Velocity corrections are measured by the IMU and controlled by the LGC. During coasting phases of the mission, velocity corrections are not made continuously, but are initiated at predetermined checkpoints.

The LGC memory consists of an erasable and a fixed magnetic core memory with a combined capacity of 25,600 sixteen bit words. The erasable memory is a coincident-current ferrite core array with a total capacity of 1,024 words, and is characterized by destructive readout. Fixed memory consists of six magnetic core rope modules, each with a capacity of 4,096 words, with a total capacity of 24,576 words. Readout from fixed memory is nondestructive. The logical operations of the LGC are mechanized using micrologic elements, where the necessary resistors and transistors are diffused into single silicon wafers. One complete NOR gate, which is the basic building block for all LGC logic circuitry, is contained in a package the size of an aspirin tablet. Flip-flops, registers, counters, etc., are all comprised of these standard NOR elements in different wiring configurations. The LGC performs all necessary arithmetic operations by addition, adding two complete words and preparing for the next operation in approximately 24 microseconds. To subtract, the LGC adds the complement of the subtrahend. Multiplication is performed by successive additions and shifting, and division is performed by the successive addition of complements and shifting.

3-29. POWER AND SERVO ASSEMBLY.

The Power and Servo Assembly (PSA) provides a central mounting place for the majority of the N & G Subsystem amplifiers, modular electronic components, and power supplies. The PSA consists of the following subassemblies: gimbal servo align and power amplifiers; gyro and accelerometer amplifiers and electronics; CDU electronics; power diodes and signal conditioners; and power supplies. A cold plate, through which water-glycol coolant from the Environmental Control Subsystem flows, is mounted under the PSA subassemblies to dissipate heat.

3-30. STABILIZATION AND CONTROL SUBSYSTEM.

The Stabilization and Control Subsystem (SCS) is divided into two major sections: control electronics section (CES) and abort guidance section (AGS). The CES consists of a control panel, an attitude and translation control assembly (ATCA), a descent engine control assembly (DECA), two gimbal drive actuator assemblies (GDA), two attitude controller assemblies (ACA), two translation controller assemblies (TCA), and two rate gyro assemblies (RGA). The AGS consists of an attitude reference assembly (ARA), an abort programmer assembly (APA), a control panel, and an in-flight monitor assembly (IFMA).

3-31. CONTROL ELECTRONICS SECTION. (See figure 3-6.)

The CES converts attitude error, rate commands, or translation command signals generated in the Navigation and Guidance (N & G) Subsystem, the AGS, and manually operated cockpit controls into pulse-ratio-modulated pulsed or full-on signals for firing the thrust chambers in the Reaction Control Subsystem (RCS). Rate and attitude error signals from the CES are routed to displays; the logic circuits within the CES route the signals to the appropriate thrusters. In addition, the CES conditions on-off commands for the ascent and descent engines, and routes automatic and manual throttle commands to the descent engine. Trim control of the gimballed descent engine is also provided to assure that the thrust vector operates through the LEM vehicle's center of gravity.

3-32. Attitude Control. Five basic modes of attitude control are provided by the CES and are selectable on the control panel. (See figure 3-2.) They are: automatic mode, attitude hold mode, rate command mode, pulse mode and direct mode. The pulse and direct modes are selectable on an individual axis basis.

3-33. Automatic Mode. The automatic mode provides fully automatic attitude control. When this mode is selected, attitude error signal from the N & G Subsystem or the AGS are applied to the CES. The attitude error signals are routed to the ATCA, where they are combined with angular rate signals from the RGA. These signals are then demodulated and applied through deadband circuits, logic circuits, pulse modulators, and jet driver amplifiers to fire the appropriate RCS thrust chambers. Pitch and roll signals are also routed to the DECA for automatic trim control of the descent engine after demodulation. The selectable deadband feature provides a wide and narrow deadband (MAX and MIN). The wide deadband is used during coasting phases of the mission to conserve fuel; the narrow deadband is used when accurate thrust vector control is required. The signals applied to the logic circuit are processed and directed to the appropriate pulse modulators and jet driver amplifiers. The pulse modulators and jet drivers generate the fire pulses that actuate the RCS thrust chambers. The descent engine control commands and translation control commands from the N & G Subsystem are routed through the control panel to CES for automatic descent. Also, ascent engine on-off commands are routed through the control panel to the ascent engine during ascent.

3-34. Attitude Hold Mode. The attitude hold mode is the primary attitude control mode for the hover, landing, and docking phases of the mission. When this mode is selected, the pilot manually commands attitude change rates by displacing the attitude controller from the neutral position. When the controller is in neutral, the error signals from the N & G Subsystem or AGS causes the LEM to hold attitude.

3-35. Rate Command Mode. The rate command mode functions in the same manner as the attitude hold mode except that attitude hold is not provided when the attitude controller is in the neutral position.

3-36. Pulse Mode. The pulse mode is an open loop attitude control mode. In this mode minimum impulse attitude changes can be made in any axis using the attitude controller.

3-37. Direct Mode. The direct mode is also an open loop attitude control mode. In this mode two thruster attitude changes can be made in any axis using the attitude controller.

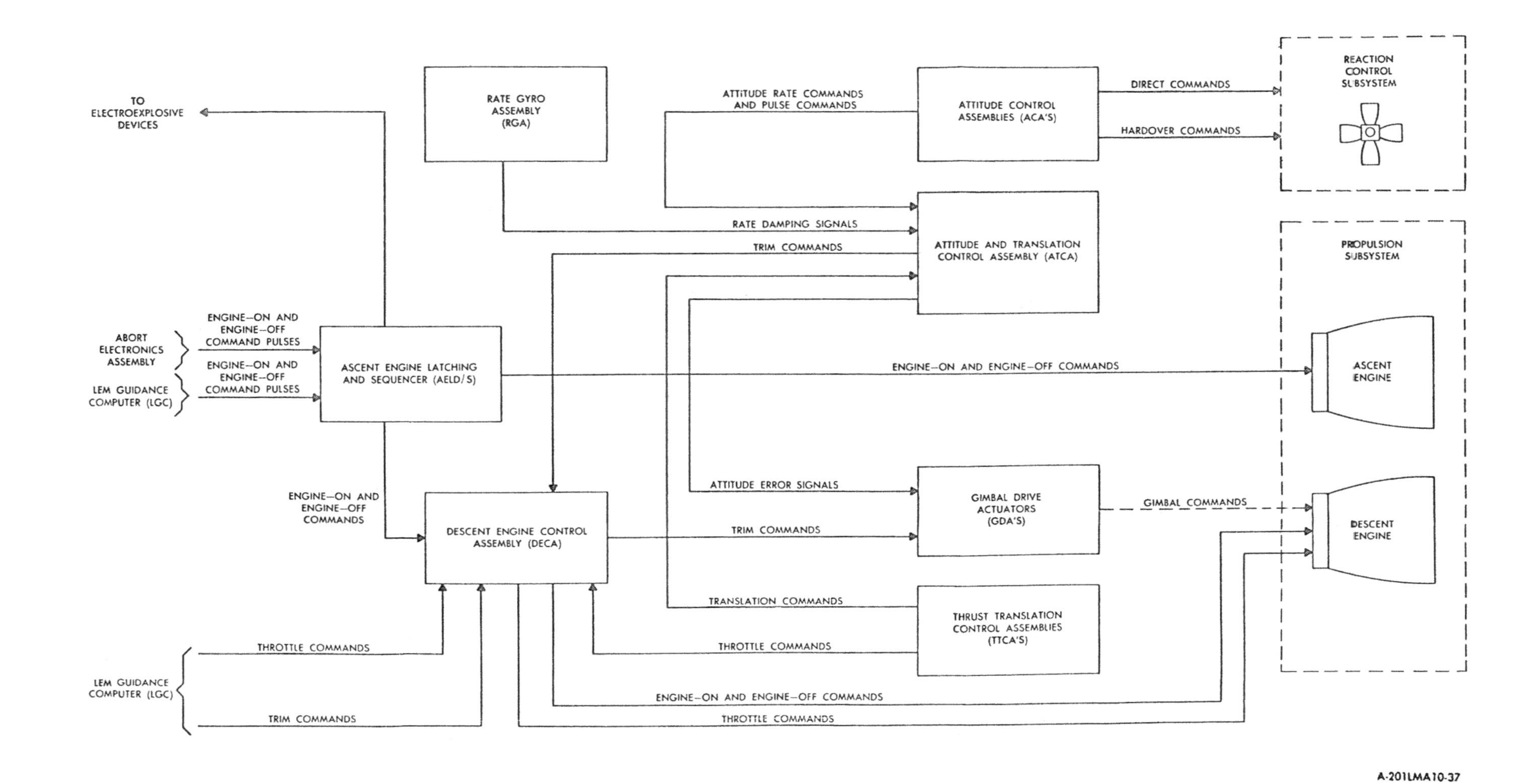

Figure 3-6. **Control Electronics Section Block Diagram**

3-38. Manual Override. Manual override can be activated in any mode by displacing the attitude controller to its full travel. This provides four thruster operation through the secondary solenoids of the RCS thrust chambers which are directly wired to switches in the attitude controller for maximum reliability.

3-39. Translation Control. Both automatic and manual translation control using the reaction jets are provided. Automatic translation control consists of on-off commands from the N & G Subsystem for the x-axis only. This is used for small velocity changes and out-of-plane correction during the mission. Manual translation control is provided in all three axes for docking and for small corrections during hover and landing. Both pulse and on-off type of manual translation control are provided. Detent switches in the translation controller insert the proper voltage level into the logic for direct operation of the RCS jets.

3-40. ABORT GUIDANCE SECTION (See figure 3-7.)

Three major functions are provided by this section: abort capabilities from any point in power descent or ascent; attitude reference for vehicle stabilization; and capabilities for ascent from the lunar surface. The AGS generates LEM attitude error signal that, after passing through the CES, actuates appropriate RCS thrusters. Additional functions are provided in the AGS by the IFMA. These functions are: monitoring the spin motor rotation of each gyro in the RGA; stimulating the SCS for OBCE (part of Instrumentation Subsystem) tests; and providing unique signal conditioning for SCS's signals to be telemetered. The AGS has three modes: warmup, alignment, and inertial reference.

3-41. Warm-Up Mode. The warm-up mode provides 28-volts d-c power to the AGS to prepare the AGS for operation 30 minutes after entering this mode.

3-42. Alignment Mode. The alignment mode has three sub-modes: IMU alignment, orbital alignment, and lunar alignment of the AGS. The IMU alignment is accomplished by aligning the AGS euler angles to the IMU gimbal angles. Oribital alignment is used if the N&G Subsystem malfunctions and the LEM vehicle is in lunar orbit for such extended periods that the subsystems must be shutdown to conserve power. Oribital alignment is accomplished by manually achieving the proper LEM vehicle attitude with respect to the Command/Service module's orbit, and manually pitching the LEM vehicle to a predetermined pitch euler angle. Lunar alignment is used in preparation for ascent from the lunar surface if the N & G Subsystem has malfunctioned. Lunar alignment is accomplished by manually setting the yaw euler angle to the same value obtained from the AGS control panel after lunar landing. The pitch and roll euler angles are determined by local vertical sensors.

3-43. Inertial Reference Mode. The inerial reference mode has four sub-modes: followup, abort attitude hold, automatic, and manual. The followup sub-mode permits the AGS to follow LEM vehicle motion to prevent transients when control is switched from the N&G Subsystem to the AGS. (LEM Attitude error signal equal zero). zero). When the AGS is selected as the inertial reference, the LEM vehicle enters the abort attitude hold sub-mode and maintains the last attitude it held prior to the switchover. The automatic sub-mode provides automatic preprogrammed abort trajectories from any point in the lunar descent or ascent, or a preprogrammed ascent trajectory from the lunar surface. In manual sub-mode, the capability is provided to allow the astronaut to accurately control the LEM's attitude and provide manual abort capabilities.

3-44. REACTION CONTROL SUBSYSTEM.

The Reaction Control Subsystem (RCS) provides rocket thrust impulses that stabilize the LEM vehicle during descent and ascent and control the vehicle attitude and translation about or along all axes during hover, rendezvous, and docking maneuvers. The RCS consists basically of 16 thrust chambers supplied by two separate and independent propellant feed and pressurization sections. The thrust chambers are mounted in clusters of four on outriggers that are arranged diametrically opposite and equally spaced around the LEM ascent stage. In each cluster, two thrust chambers are mounted on a vertical axis, facing in opposite directions; the other two are spaced 90 degrees apart, parallel to the vehicle's y and z axes. Two of the thrust chambers in each cluster are supplied by one propellant feed section; the remaining two are supplied by the other independent feed section. Both systems are normally in simultaneous operation, but the arrangement is such that complete control in all axes is still possible, despite a failure in either propellant supply. A schematic diagram of the RCS is shown in figure 3-8. For a view of the RCS installation in the vehicle, see figure 3-9.

3-45. PROPELLANTS.

The RCS utilizes hypergolic propellants consisting of a 50-50 fuel mixture of hydrazine (N_2H_4) and unsymmetrical dimethylhydrazine (UDMH), with nitrogen tetroxide (N_2O_4) as the oxidizer. The mixture ratio of oxidizer to fuel is 2 to 1, by weight; both ignite spontaneously upon contact with each other. This same propellant combination is also used in the LEM ascent and descent propulsion subsystems.

3-46. PROPELLANT FEED AND PRESSURIZATION SECTIONS.

There are dual propellant feed and pressurization sections, commonly referred to as system A and system B. Each propellant feed section has two cylindrical tanks, one for fuel and one for oxidizer, and these propellants are contained within positive expulsion bladders that are supported by standpipes running lengthwise through each tank. Both tanks are pressurized by a common supply of helium, which acts upon the bladders to expel the propellants. The helium is stored at a pressure of 3,000 psi in a spherical tank and two parallel, explosively operated squib valves seal off this supply until they are fired prior to the initial RCS start. Downstream of the squib valves, the helium passes through a filter and the supply line splits into parallel legs, each containing two pressure regulators in series that step down the helium pressure to approximately 179 psi. Solenoid valves are located in each leg immediately upstream of the pressure regulators, and are operated such that only one regulator leg is open at a time. Should the normally open leg malfunction, it can be closed off and the parallel leg opened.

The helium supply then branches into oxidizer tank and fuel tank pressurization lines, with a quadruple check valve in each. There is a relief valve in the helium line to each propellant tank to prevent system overpressurization if the pressure regulator set should fail completely. Even with such failure, however, the relief pressure (set at approximately 250 psi) permits uninterrupted tank pressurization at this setting for continued RCS operation. A burst disk and filter assembly is located immediately upstream of each relief valve; the burst disk ensures against possible leakage during normal operation.

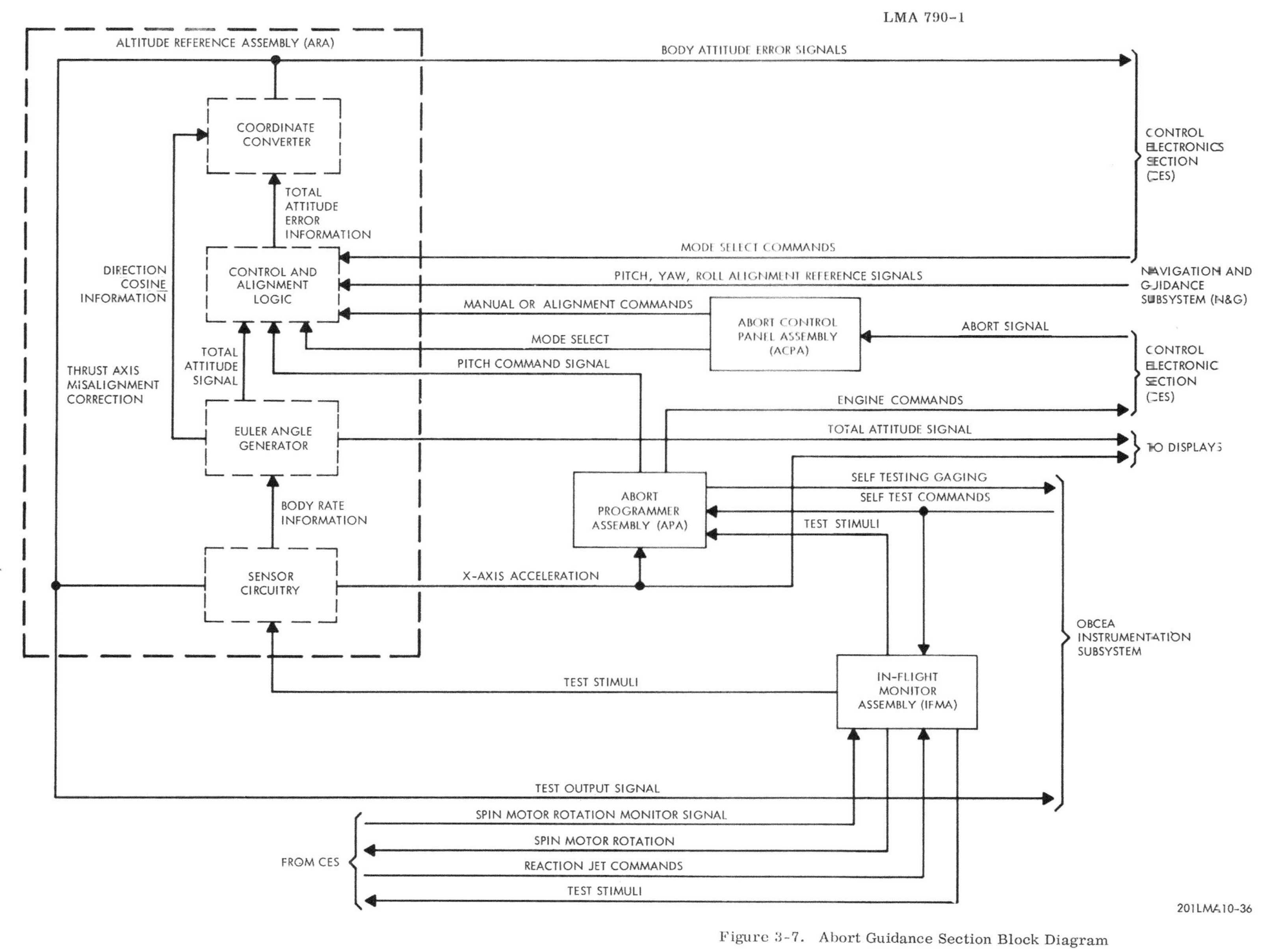

Figure 3-7. Abort Guidance Section Block Diagram

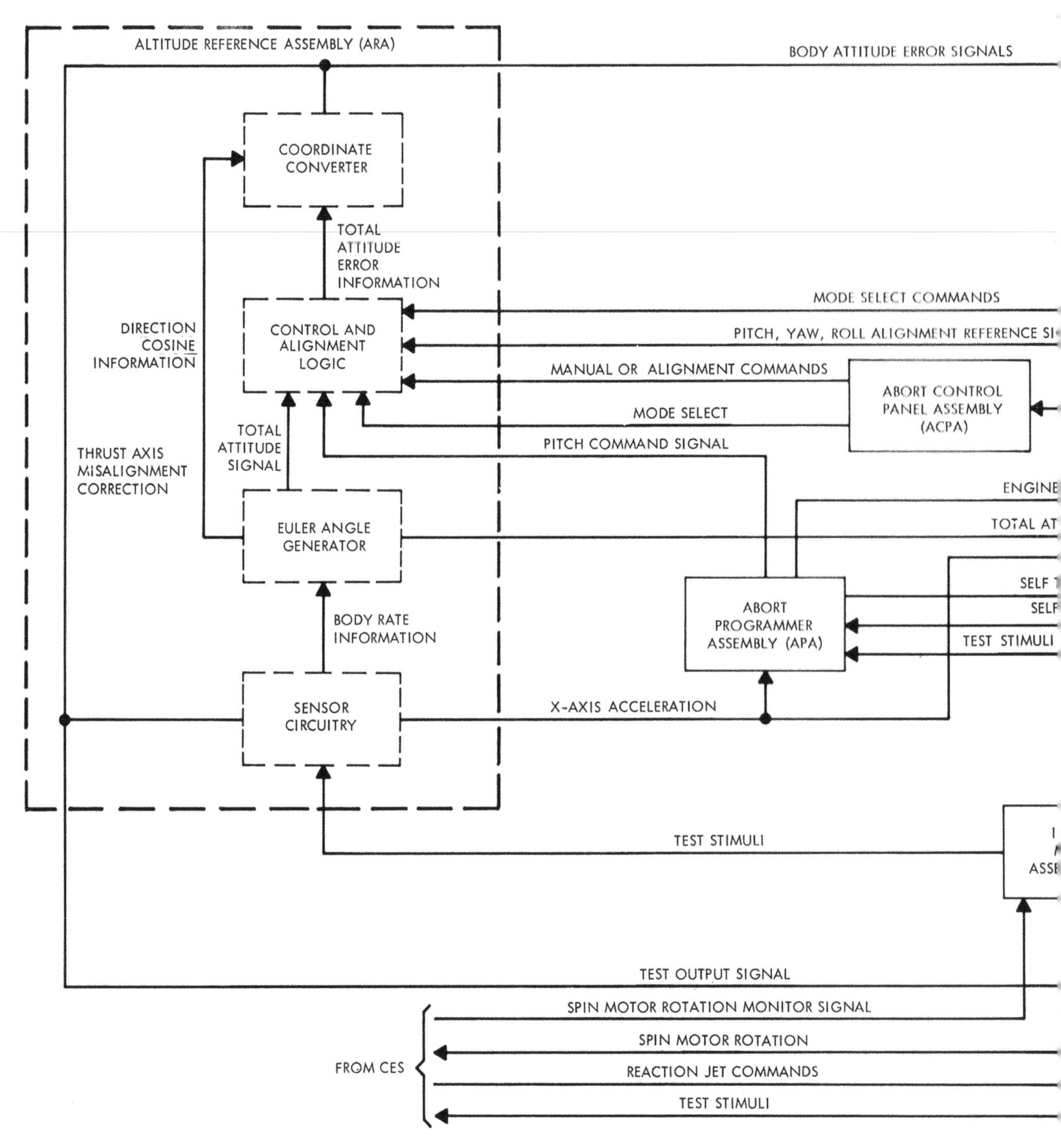

Figure 3-7. Abort Gu

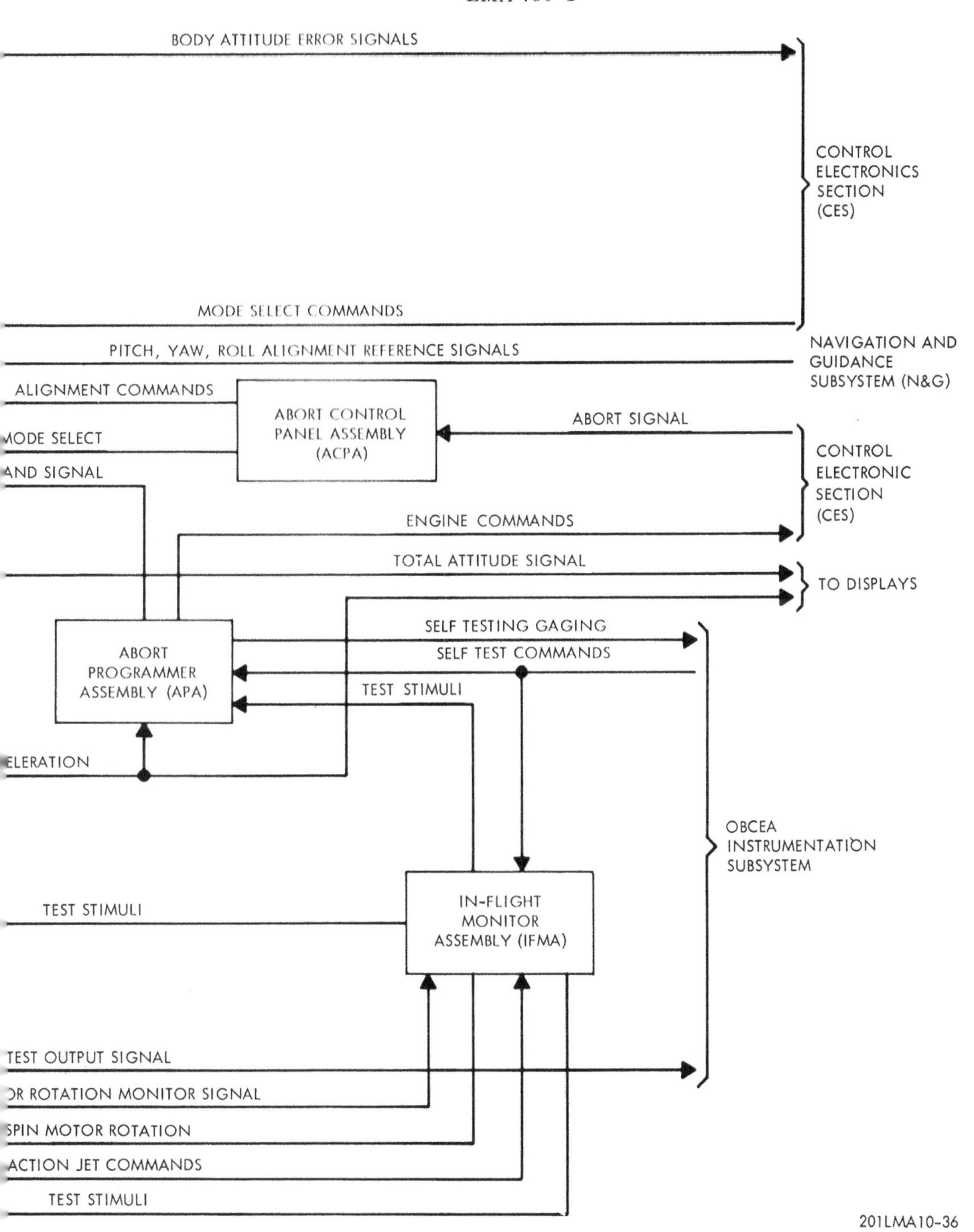

Figure 3-7. Abort Guidance Section Block Diagram

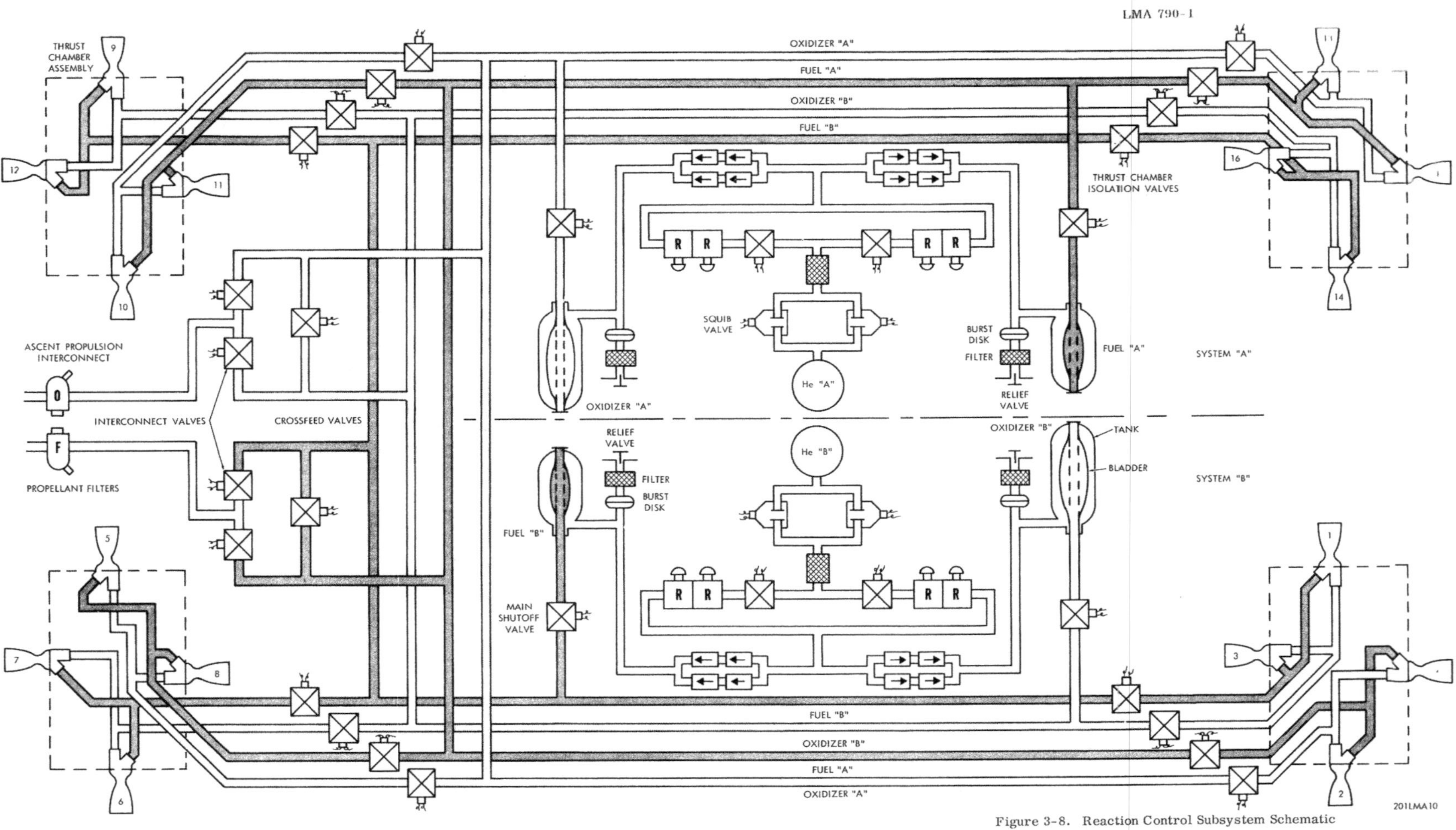

Figure 3-8. Reaction Control Subsystem Schematic

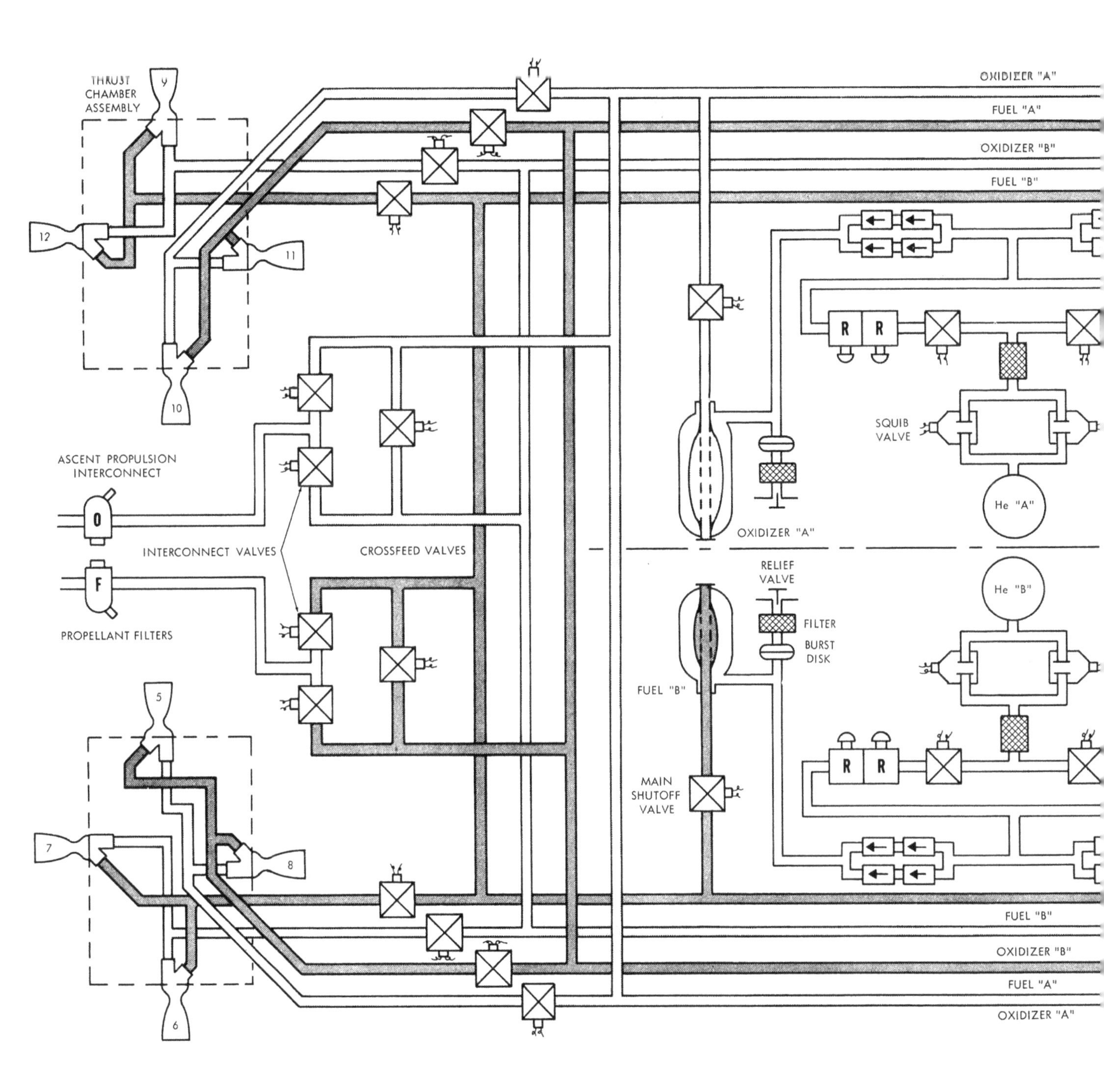
THRUST
CHAMBER
ASSEMBLY
OXIDIZER "A"
FUEL "A"
OXIDIZER "B"
FUEL "B"
9
12
11
10
R
R
SQUIB
VALVE
He "A"
ASCENT PROPULSION
INTERCONNECT
O
F
INTERCONNECT VALVES
CROSSFEED VALVES
PROPELLANT FILTERS
OXIDIZER "A"
RELIEF
VALVE
He "B"
FILTER
BURST
DISK
FUEL "B"
5
MAIN
SHUTOFF
VALVE
7
8
6
FUEL "B"
OXIDIZER "B"
FUEL "A"
OXIDIZER "A"

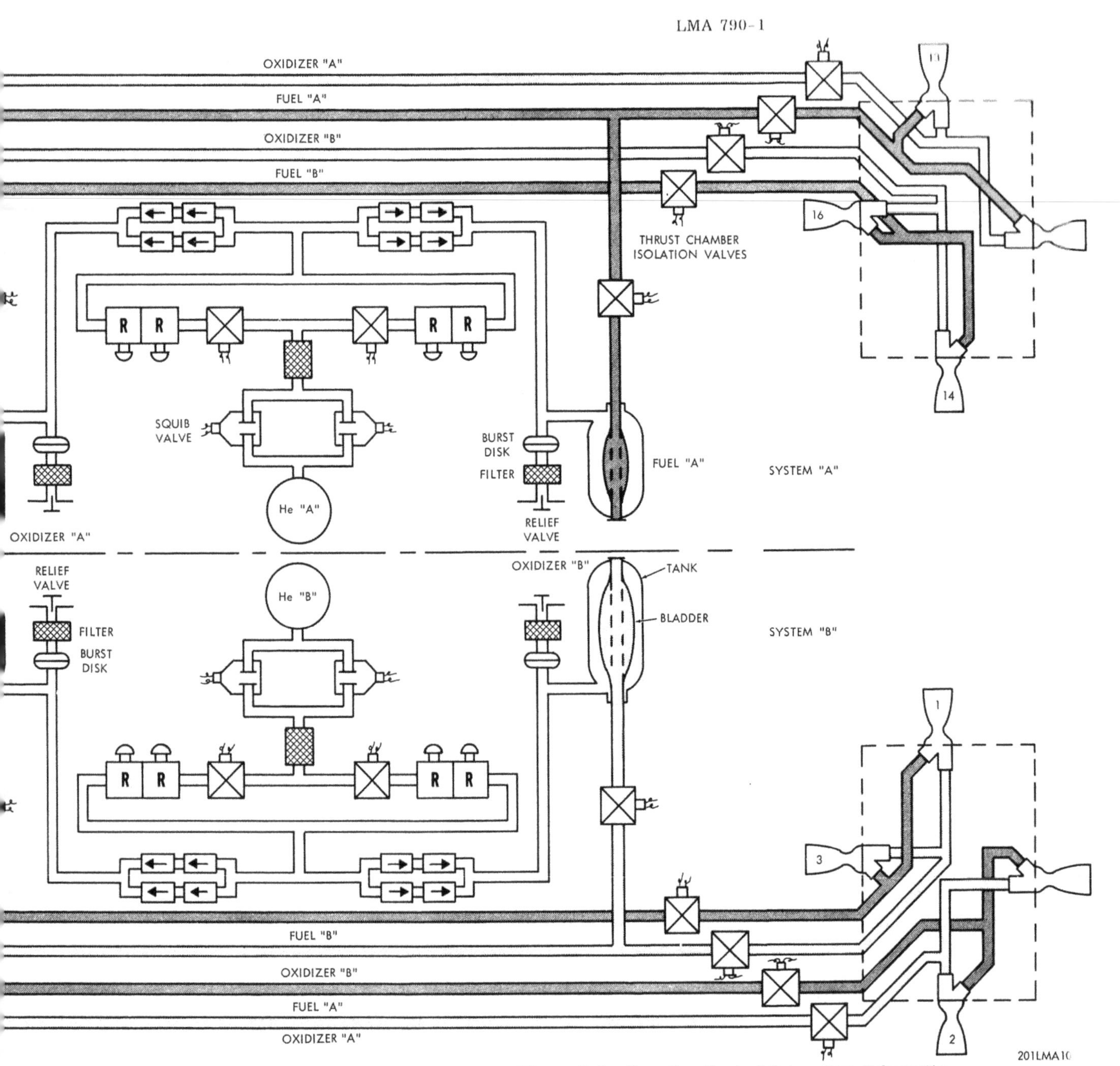

Figure 3-8. Reaction Control Subsystem Schematic

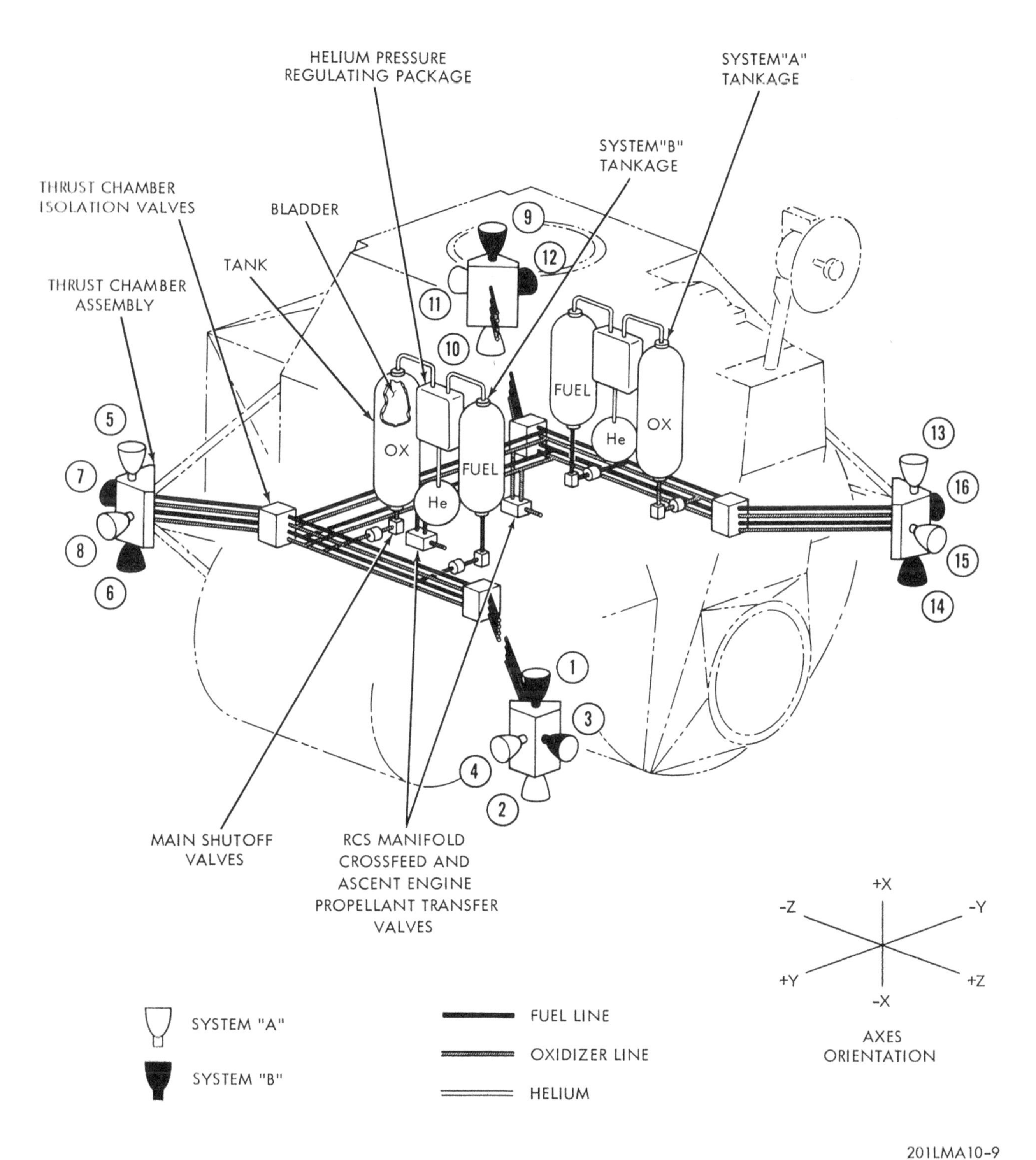

Figure 3-9. Reaction Control Subsystem Installation

As the tanks are pressurized during the initial RCS start, the fuel and oxidizer from each feed section then flow into a manifold that supplies eight of the 16 thrust chambers. Normally open solenoid valves in the outlet line of each propellant tank may be closed to isolate the feed section if a malfunction occurs. If one feed section must remain isolated in this manner, normally closed solenoid valves in a crossfeed piping arrangement between the system A and system B manifolds may be opened to provide propellant flow from the other feed section to all 16 thrust chambers. In addition, there are similar solenoid valves in feed lines connecting the RCS manifolds and the ascent propulsion propellant feed lines that provide for the transfer of ascent engine propellant to the RCS thrust chambers, if necessary.

3-47. THRUST CHAMBER ASSEMBLIES.

Each thrust chamber is a small rocket engine that develops 100 pounds of thrust and is capable of either pulse-mode or steady-state operation. The engine consists primarily of a combustion chamber and nozzle, an injector assembly, and fuel and oxidizer solenoid valves. (See figure 3-10). Fuel and oxidizer are piped through the cores of the solenoid valves, the armatures of which are normally seated on the injector inlets, closing off flow to the combustion chamber. Each solenoid has a primary coil and a separate secondary coil; the secondary coils are wired directly to the attitude hand controllers, while the primary coils receive signals from the electronic subsystems for various modes of RCS operation, as discussed in paragraphs 3-48 through 3-50. When the "engine-on" signal is received, fuel and oxidizer solenoids are energized simultaneously, lifting the armatures from their seats to permit flow to the injector orifices. These orifices are drilled in doublets at angles to each other so that the emerging fuel and oxidizer streams impinge. Normally open solenoid valves in the propellant manifolds just upstream of each thrust chamber assembly may be closed to isolate the engines, should the cluster malfunction or be damaged.

3-48. RCS OPERATIONAL MODES.

The RCS is initially fired for LEM separation from the Command/Service modules, and may be operated by the astronauts at any time thereafter, either for attitude or for translation control.

3-49. Attitude Control. Five modes of attitude control are provided: one is fully automatic, two are semiautomatic (attitude hold and rate command modes), and two are manual (direct and pulse modes). In fully automatic and semiautomatic operation, control commands originate in either the Navigation and Guidance Subsystem or the abort guidance section of the Stabilization and Control Subsystem, from whence they are processed through the Control Electronics Section (CES) of the Stabilization and Control Subsystem in order to establish which of the 16 thrust chambers shall be fired, when, and for how long. Outputs from the CES are timed voltage impulses that energize the primary coils in the propellant solenoid valves on the appropriate thrust chambers. The fully automatic mode is normally used to provide attitude control during all phases of the mission, except hover, landing, and docking. It is possible, however, to select manual control in one or two axes and retain automatic control in the other axis during all mission phases.

Semiautomatic control is used during the attitude hold and rate command modes of operation. The attitude hold mode combines automatic attitude hold with manual control about each axis through the astronaut's hand attitude controller, and is used primarily to control the LEM during hover, landing, and docking phases of the

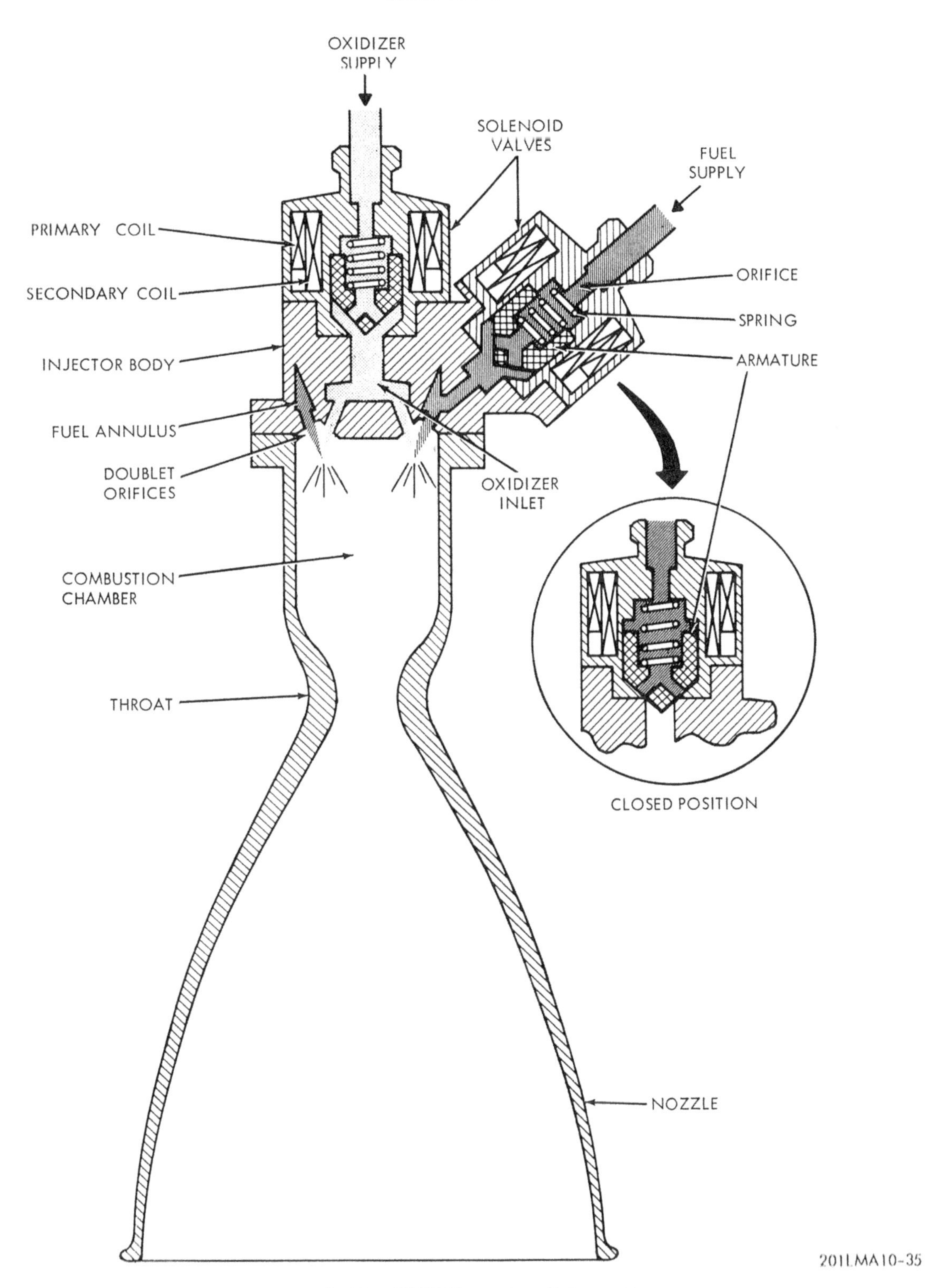

Figure 3-10. RCS Thruster Schematic

mission. The rate command mode is essentially similar to attitude hold, but lacks the automatic attitude hold feature.

In the manual modes of operation, all control commands originate from the hand attitude controller, including manual control of the duration of thrust. Pulse mode operation is obtained by normal movement of the hand attitude controller throughout its range of travel; a pulse frequency of about two cycles per second is selected and processed through the CES to energize the thrust chamber primary coils. The direct mode of operation is obtained by moving the controller to the full extent of its travel to actuate limit switches that are wired directly to the thrust chamber secondary coils. This latter type of control bypasses the electronics subsystems entirely and provides a rapid override.

3-50. Translational Control. Manual and automatic modes of translational control are provided in all three axes. Manual control commands originate in the hand translational controller; as the controller is moved throughout its range of travel, control commands are processed through the CES and routed to the thrust chamber primary coils as fixed pulse-rate signals. Continued movement of the controller to the full extent of its travel actuates limit switches; in this instance, however, the control commands are then processed through the CES and transmitted as pure "on-off" signals to the thrust chamber primary coils. Automatic translational commands originate in the Navigation and Guidance Subsystem and are likewise processed through the CES as pure "on-off" signals and routed to the primary coils.

3-51. LEM PROPULSION.

The LEM utilizes separate descent and ascent propulsion subsystems, each of which is complete and independent of the other, and consists basically of a liquid propellant rocket engine with its propellant storage, pressurization, and feed components. The Descent Propulsion Subsystem is contained wholly within the vehicle descent stage and utilizes a throttleable, gimballed engine that is first fired to inject the LEM into the descent transfer orbit and subsequently used in the final descent trajectory as a retrorocket to control the rate of descent and also to enable the LEM to hover and move horizontally. The Ascent Propulsion Subsystem is installed entirely within the ascent stage, but uses a fixed, constant-thrust engine to generate the thrust sufficient to launch the ascent stage from the lunar surface and place it in orbit. The ascent engine is also capable of providing gross orbit adjustments necessary for successful rendezvous with the Command/Service modules.

The Control Electronics Section of the Stabilization and Control Subsystem provides automatic "on-off" commands for both engines and initiates thrust magnitude and gimbal actuator commands for the descent engine. A manual override for engine control is also available to the crew.

The propellant used in both subsystems is a 50-50 fuel mixture of hydrazine and unsymmetrical dimethylhydrazine, with nitrogen tetroxide as the oxidizer, the same as is used in the Reaction Control Subsystem. The mixture ratio of oxidizer to fuel for propulsion, however, is 1.6 to 1 by weight. In both stages, the propellants are supplied from baffled tanks, with helium as the tank pressurant.

The Descent Propulsion Subsystem is described in the paragraphs immediately following; the Ascent Propulsion Subsystem is covered in paragraphs 3-56 and subsequent.

3-52. DESCENT PROPULSION SUBSYSTEM.

The Descent Propulsion Subsystem consists of two fuel and two oxidizer tanks with associated valving, piping, and pressurization components, and a single, deep-throttling engine that develops 10,500 pounds of thrust in a vacuum at full throttle and 1,050 pounds at minimum thrust. The engine is mounted in the center compartment of the descent stage (see figure 2-4), and is pivoted at the throat of the combustion chamber on a gimbal ring that is an integral portion of the engine assembly. The ring is suspended in the structure, and is pivoted in a line 90 degrees opposite to the engine pivots, so that the engine is capable of being gimballed ±6 degrees in any direction by vehicle-mounted gimbal actuators in order to provide trim control in the pitch and roll axes during powered descent. The descent engine consists primarily of an ablative combustion chamber, an injector, throttling and isolation valve packages, and a radiation-cooled nozzle extension. The nozzle extension is designed to crush, should it contact the lunar surface upon landing. The engine also has instrumentation sensors to measure fuel and oxidizer inlet pressures and temperatures, injector inlet pressures, thrust chamber pressures, valve positions, vibration, and exterior surface temperatures.

The 10:1 throttling range requirement for the descent engine has fostered a development effort aimed at maintaining uniform injector spray patterns over the entire range of propellant flow rates in order to assure high performance and stable combustion at all thrust settings. This effort is currently the subject of a parallel study in which an engine that employs the injection of an inert gas (helium) into the propellant manifolds to sustain propellant injection velocities is being tested and evaluated in competition with a mechanically throttled engine in order to select the optimum throttling technique. One design will be chosen subsequent to this writing. For the purposes of this manual, however, both are described in subsequent paragraphs.

3-53. Descent Propulsion Propellant Feed and Pressurization Sections. The propellant feed and pressurization sections include the tankage and helium supply assemblies necessary for the delivery of fuel and oxidizer to the rocket engine valve packages. Because two different engine designs are being considered, two schematics of the propellant feed and pressurization sections are presented, showing the applicable modifications. The sections used with the mechanically throttled engine are shown in figure 3-11; an alternate design for the helium-injection-throttled engine is shown in figure 3-12.

The descent propulsion propellant tanks are pressurized by a supply of helium that is stored at approximately 3,500 psi in two, interconnected spherical vessels and piped through a series of valves and pressure reducing regulators. Both storage vessels discharge into a common manifold; a normally closed, explosive-operated squib valve isolates the helium supply prior to start. Downstream of this valve, the gas passes through a filter that retains any debris produced by firing the valve, after which the supply line divides into two flow paths, with a normally open solenoid valve and two pressure regulators in series in each leg. The solenoid valves are closed during the coast periods of descent to prevent possible tank overpressurization. Regulators 3 and 4 are set to a slightly lower pressure than those in the parallel leg; normal operation is for 3 and 4 to remain locked up while helium pressure is regulated through numbers 1 and 2. In addition, the upstream regulators 1 and 3 are set to deliver a slightly lower pressure than each of the downstream regulators. In normal operation, regulator number 2 remains fully open, sensing a demand, while number 1 regulates. Should regulator 1 fail open, control is obtained through

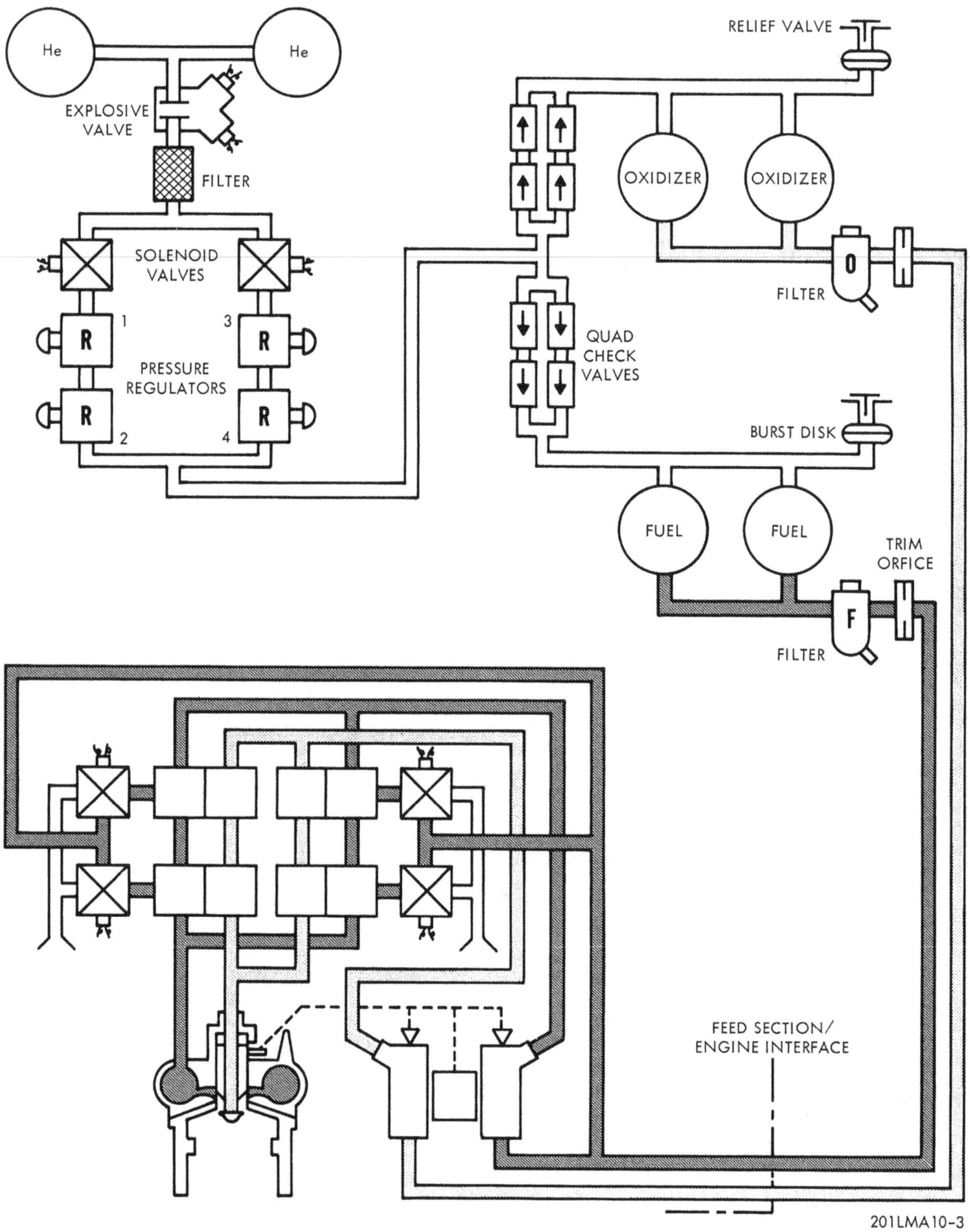

Figure 3-11. Descent Propulsion Propellant Feed and Pressurization Schematic (Mechanical Throttling)

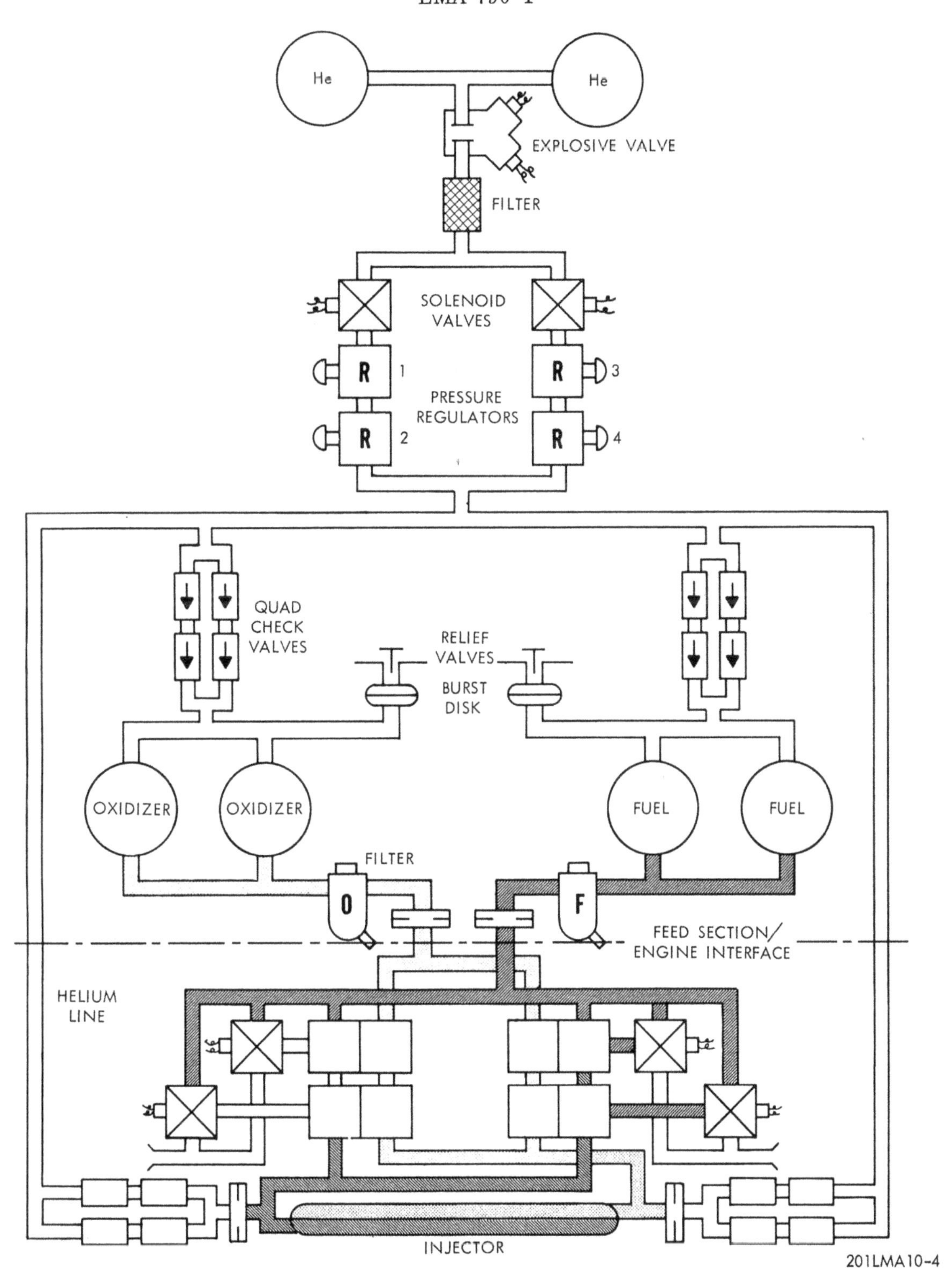

Figure 3-12. Descent Propulsion Propellant Feed and Pressurization Schematic (Helium Injection Throttling)

number 2. Should regulator 1 or 2 fail closed, control is obtained in a similar manner through regulators 3 and 4.

Downstream of the regulator assembly, the helium flow divides, passing through quadruple check valves to the fuel and oxidizer tanks. A burst disk and relief valve assembly is tapped into the helium supply line near the tank inlets to prevent dangerous tank overpressurization. The burst disk prevents possible helium leakage through a malfunctioning relief valve during normal pressurization. Additional flow lines are provided for the delivery of helium pressure to the helium-injection-throttled engine, as shown in figure 3-12.

Descent engine fuel and oxidizer are carried in four baffled tanks, two for each. Helium is introduced directly into these tanks and acts upon the surface of the fluid to expel the propellants. Each pair of tanks is manifolded into a common discharge line containing a filter and trim orifice, from whence fuel and oxidizer are piped directly to the engine.

3-54. Helium-Injection-Throttled Engine. The helium injection throttling technique involves the introduction of helium directly into the injector at low thrust settings in order to maintain propellant flows, velocities, and proper impingement angles that would ordinarily be obtained with propellant alone at full thrust. Throttling is accomplished by metering the flow of propellant through upstream bipropellant valves. A schematic of this engine valving is shown in figure 3-13.

Immediately downstream of the trim orifices at the feed section/engine interface, the fuel and oxidizer supply lines divide to provide a normal and an alternate flow path to the engine injector. Only one side operates at a time; however, switchover from the normal to the alternate leg or vice-versa may be accomplished at any time. Each flow path contains a throttling and an isolation valve, each actuated by fuel pressure metered through solenoid actuated pilot valves. Both are poppet type valves, arranged in a fuel/oxidizer pair; the poppets are actuated by a common shaft attached to an actuator piston and open and close simultaneously. When the engine start sequence is initiated with the pressurization of the propellant tanks, fuel and oxidizer pass to the engine valving. At the fire command signal, the shut-off valve solenoid is energized, feeding fuel to the actuator piston of the isolation valve. A servo valve set is included in the thrust control loop and operates in like manner to apply fluid to the throttling valve actuator. The servo valve, however, consists of separate pilot and vent valves, each with its own solenoid. Both solenoids are tied into a bang-bang control circuit, which produces an alternating switch-over mode of operation between the pilot and vent valves in order that the throttle valve is maintained in proper position at all times by fuel pressure metering through the servo valve. The throttle valve controls the flow of fuel and oxidizer to the injector at a constant mixture ratio through properly shaped valve metering contours and regulates the chamber pressure as a function of propellant flow rate.

The helium supply is obtained from the propellant tank pressurization section and is also introduced at the feed section/engine interface in two separate flow paths. The helium is then piped through quadruple on-off solenoid valves arranged in series-parallel, from whence it is injected directly into the propellant inlet manifolds of the engine injector. When the engine is throttled below 50% of full thrust, the helium valve set opens and supplies helium to the injector at a constant pressure and flow sufficient to maintain uniform propellant injection velocities at all settings down to minimum thrust. Two sets of helium quad valves are provided, one for fuel manifold

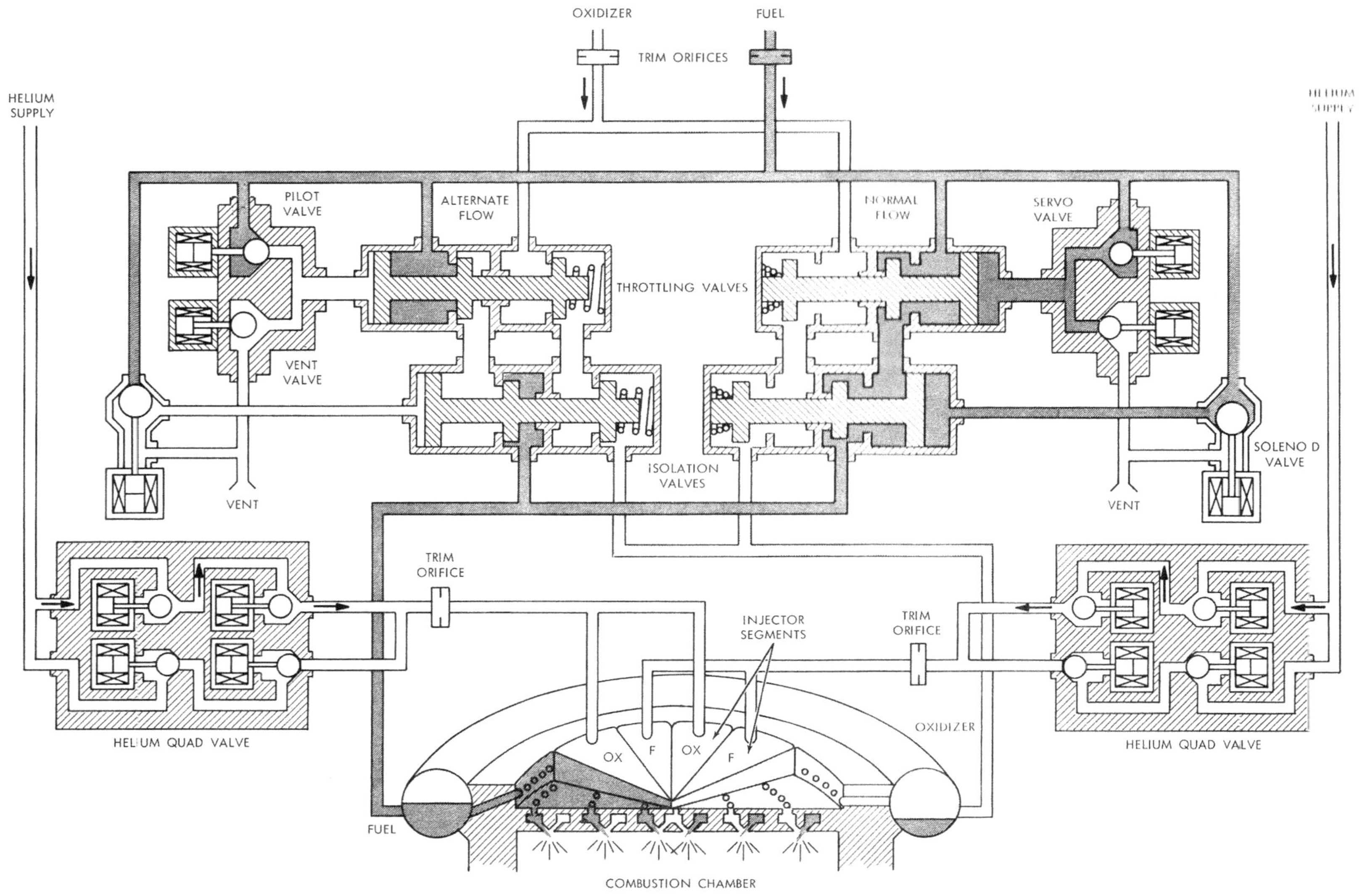

Figure 3-13. Propellant Injector and Valves
(Helium-Injection-Throttled Engine)

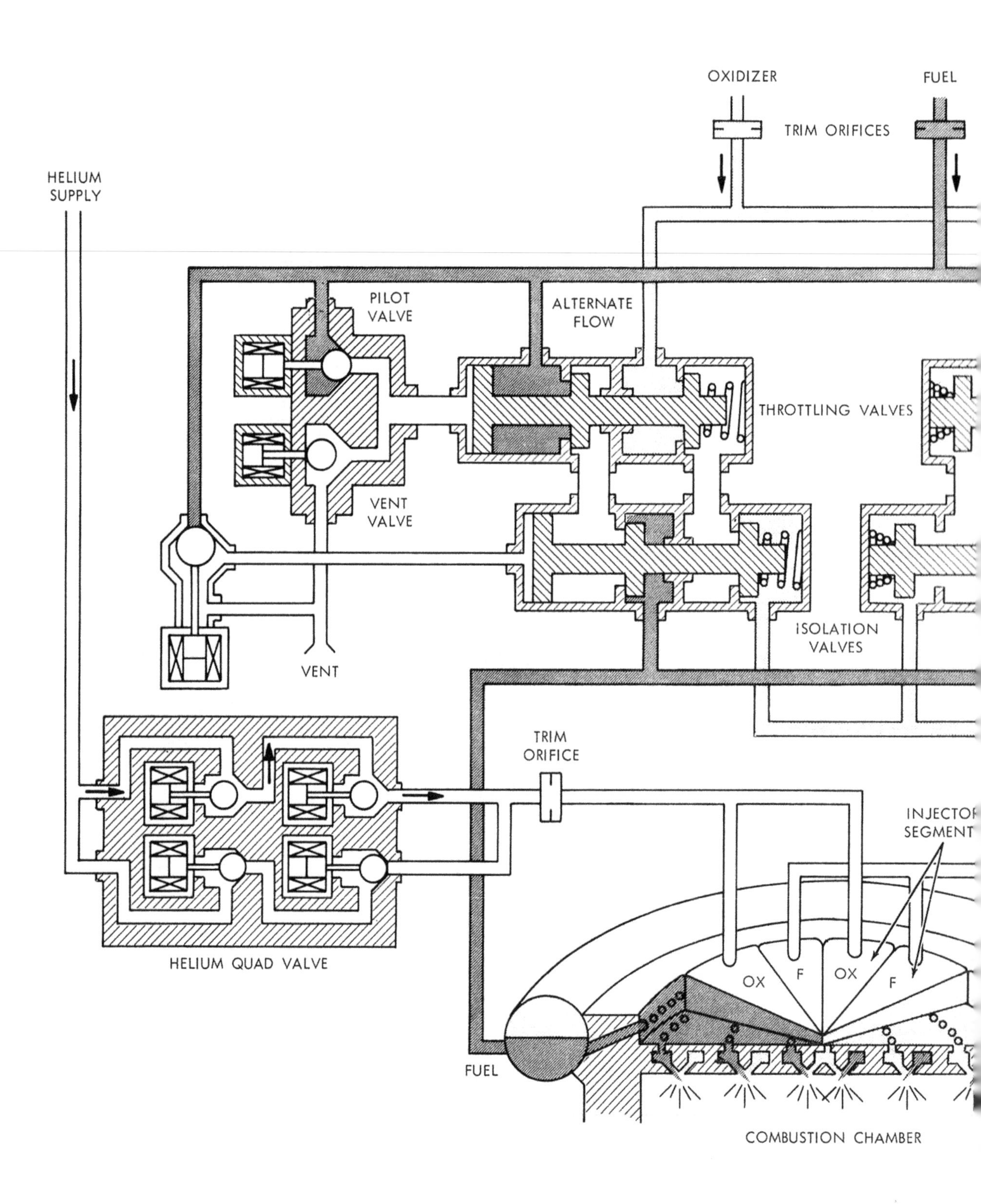
OXIDIZER
FUEL
TRIM ORIFICES
HELIUM
SUPPLY
PILOT
VALVE
ALTERNATE
FLOW
THROTTLING VALVES
VENT
VALVE
ISOLATION
VALVES
VENT
TRIM
ORIFICE
HELIUM QUAD VALVE
OX
F
OX
F
FUEL
COMBUSTION CHAMBER

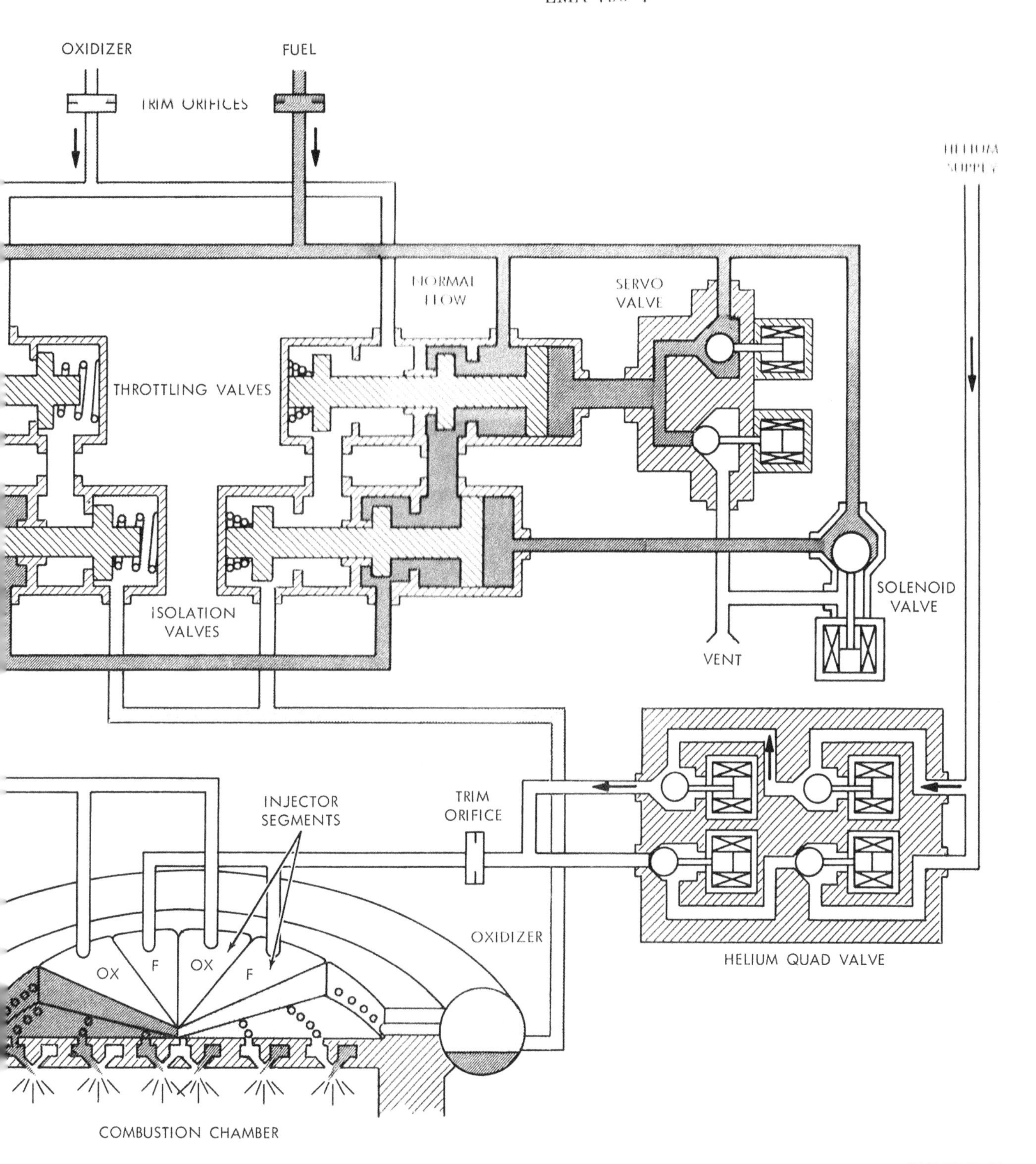

Figure 3-13. Propellant Injector and Valves (Helium-Injection-Throttled Engine)

injection and one for oxidizer manifold injection; the arrangement is such that it prevents the possibility of fuel from the injector backing up into the oxidizer manifolds, or vice versa. Within each valve itself, only one side is in operation at a time; the other flow path is redundant and switchover may be accomplished at any time.

This engine uses a fixed-area, fixed-orifice injector. The fuel and oxidizer orifices are arranged in a series of concentric rings on the injector face and these orifices are fed by propellant manifolds arranged in the form of pie-shaped segments on top of the injector body. The segments are arranged in an alternating fuel-oxidizer-fuel fashion. Fuel and oxidizer are routed through an annular manifold around the periphery of the injector assembly and directed into the segments. Helium is also injected directly into these segments and mixed with the propellant; as propellant pressure is reduced by throttling, the constant pressure of the inert gas maintains the velocity of propellant flow. The injector orifices are drilled at angles so that the fuel and oxidizer streams impinge on each other below the face of the injector.

Engine shutoff is accomplished by deenergizing the shutoff valve of the main propellant isolation valve. The vent valve of the throttling set is likewise opened and fuel pressure from the valve actuators is vented overboard into space.

3-55. Mechanically Throttled Engine. The mechanical throttling scheme utilizes variable-area, cavitating venturi flow control valves mechanically linked to a variable area injector in order to separate propellant flow control and propellant injection functions so that each may be optimized without compromising the other and to ensure that propellant flow rates are made insensitive to downstream pressure variations in the injector and combustion chamber. The injector consists basically of a faceplate and fuel manifold assembly outfitted with a coaxial oxidizer feed tube and movable metering sleeve. Oxidizer flows through the center tube and out between a fixed pintle and the bottom of the sleeve, while the fuel orifice is an annular opening between the sleeve contour and injector face. The design of the sleeve is such that the propellant orifices increase in area as the sleeve is moved upward, away from the fixed pintle. The areas of the throats in the flow control valves are regulated by close-tolerance, contoured metering pintles that are linked directly to the injector sleeve. The linkage is, in turn, operated by a single actuator. Thrust is thus regulated by movement of the actuator to simultaneously adjust the pintles in the flow control valves and the movable sleeve in the injector. The fuel and oxidizer are thereby injected at velocities and angles compatible with variations in propellant weight flow rates. A schematic of the engine injector and valving is shown in figure 3-14.

At maximum thrust, this engine operates as a conventional, pressure-fed rocket. As the engine is throttled, injector orifice areas are decreased so that injection velocities and impingement angles of fuel and oxidizer are maintained at optimum conditions for combustion efficiency. At the same time, the pintles in the flow control valves are moved to decrease the flow control area in the venturis so that the pressure drop across the valve balances out the differential between engine inlet and injector inlet pressures. At approximately 70% of maximum thrust, cavitation commences in the valve throats. From this level down to minimum thrust, the valve functions as a cavitating venturi. Once cavitation begins, the propellant metering function is entirely removed from the injector; flow is controlled entirely by the cavitating venturi valves. The mechanical linkage connecting the valve pintles and injector sleeve is pivoted about a fulcrum attached to the injector body; the throttle control is an electromechanical linear servo actuator with parallel d-c motors that operate to position the linkage in response to electrical input signals. At maximum thrust, the actuator positions the linkage to set the flow control valves and injector apertures to the full open position. As thrust is reduced, the actuator reduces the flow at the control valves and moves the injector sleeve to reduce the apertures. The flow control valve is also equipped with a propellant temperature compensation actuator to operate a vernier adjustment in the linkage connecting the fuel and oxidizer pintles. The adjustment changes the relative rate of travel of the pintles for off-nominal conditions in order to maintain propellant mixture ratio control over the entire matrix of thrust level, propellant temperature, and propellant pressure differentials.

Fuel and oxidizer are introduced to the engine through flexible inlet lines located near the gimbal ring at the engine throat, from whence they are piped directly into the flow control valves. After flowing through the venturis, the propellants pass to a series-parallel shutoff valve assembly. The shutoff valves are fuel-pressure-actuated ball valves; fuel is introduced to the valve actuators through solenoid operated pilot valves, all of which are energized simultaneously during engine startup. During the start, the solenoids unseat the caged balls from the actuator inlet ports and seat them against the overboard vent ports. Fuel enters the actuator cavities; the actuator pistons connected to rack-and-pinion linkages twist the ball valves 90 degrees to the open position, permitting flow to the injector. The series-parallel redundancy in the valve arrangement provides for positive start and cutoff. During shutdown, the solenoids are de-energized, opening the vent ports. The spring-loaded actuators close the shutoff valves and residual actuating fuel is vented overboard into space.

3-56. ASCENT PROPULSION SUBSYSTEM.

The Ascent Propulsion Subsystem utilizes a fixed, constant-thrust rocket engine installed along the centerline of the ascent stage midsection, and includes the associated propellant feed, tankage, and pressurization components. The engine develops 3,500 pounds of thrust in a vacuum, sufficient to launch the ascent stage from the lunar surface and place it in orbit. Two main propellant tanks are used, one for fuel and one for oxidizer, installed on either side of the ascent stage structure. The propellant feed section in this subsystem includes provisions for fuel and oxidizer crossfeed to the Reaction Control Subsystem as a backup propellant supply for the latter. For a view of the engine and tankage locations in the ascent stage structure, see figure 2-3.

3-57. Ascent Propulsion Propellant Feed and Pressurization Sections. The ascent propulsion propellant feed and pressurization sections consist of helium and propellant supply assemblies functionally similar to those used in the Descent Propulsion Subsystem. A schematic is shown in figure 3-15.

Helium is stored in two separate tanks; a normally closed explosive squib valve in the line immediately downstream of each tank isolates this supply until it is fired prior to the initial engine start. The helium flow then passes into two parallel regulator lines, with a filter, a normally open solenoid valve, and two pressure reducers in series in each. The upstream reducers in each line are set to a lower pressure than those downstream; the series pair in one line is also set to deliver a lower pressure than the pair in the parallel line. Normal operation is for the reducers with the lower pressure setting to lock up as the tanks are pressurized; only one of the parallel lines will operate at a time. Should either reducer in the line fail

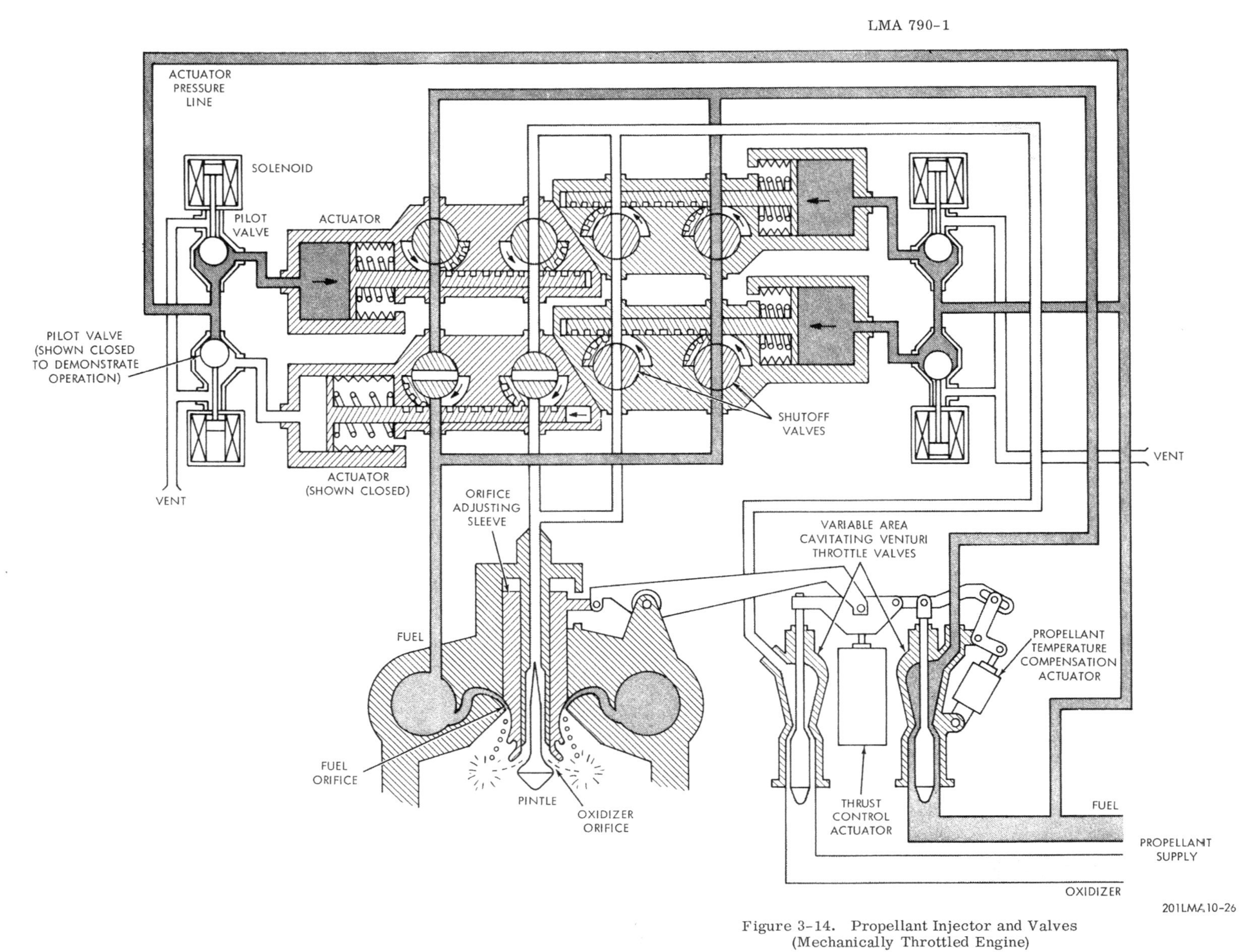

Figure 3-14. Propellant Injector and Valves
(Mechanically Throttled Engine)

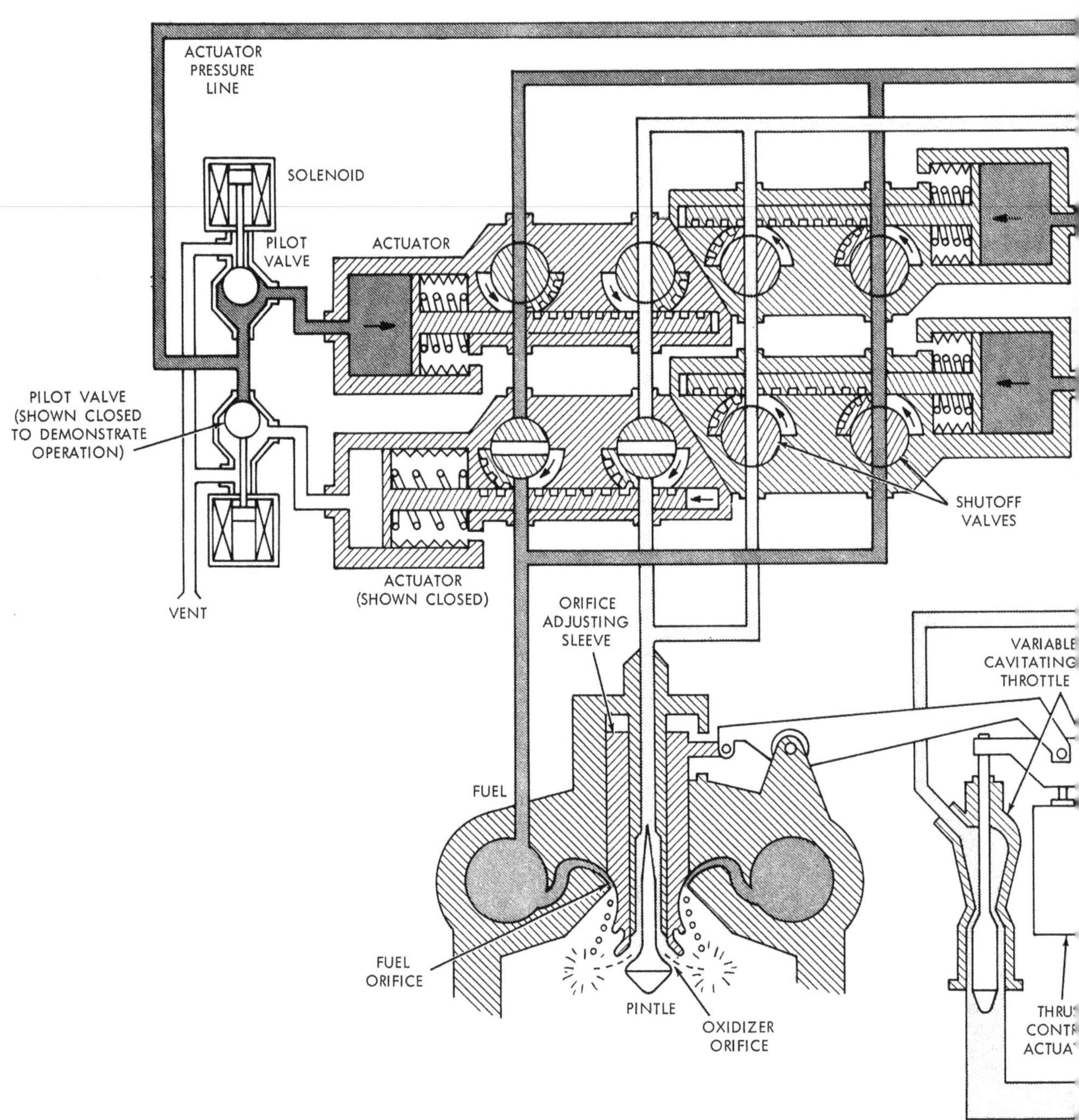

Figure 3-14. Pr
(Mechanica

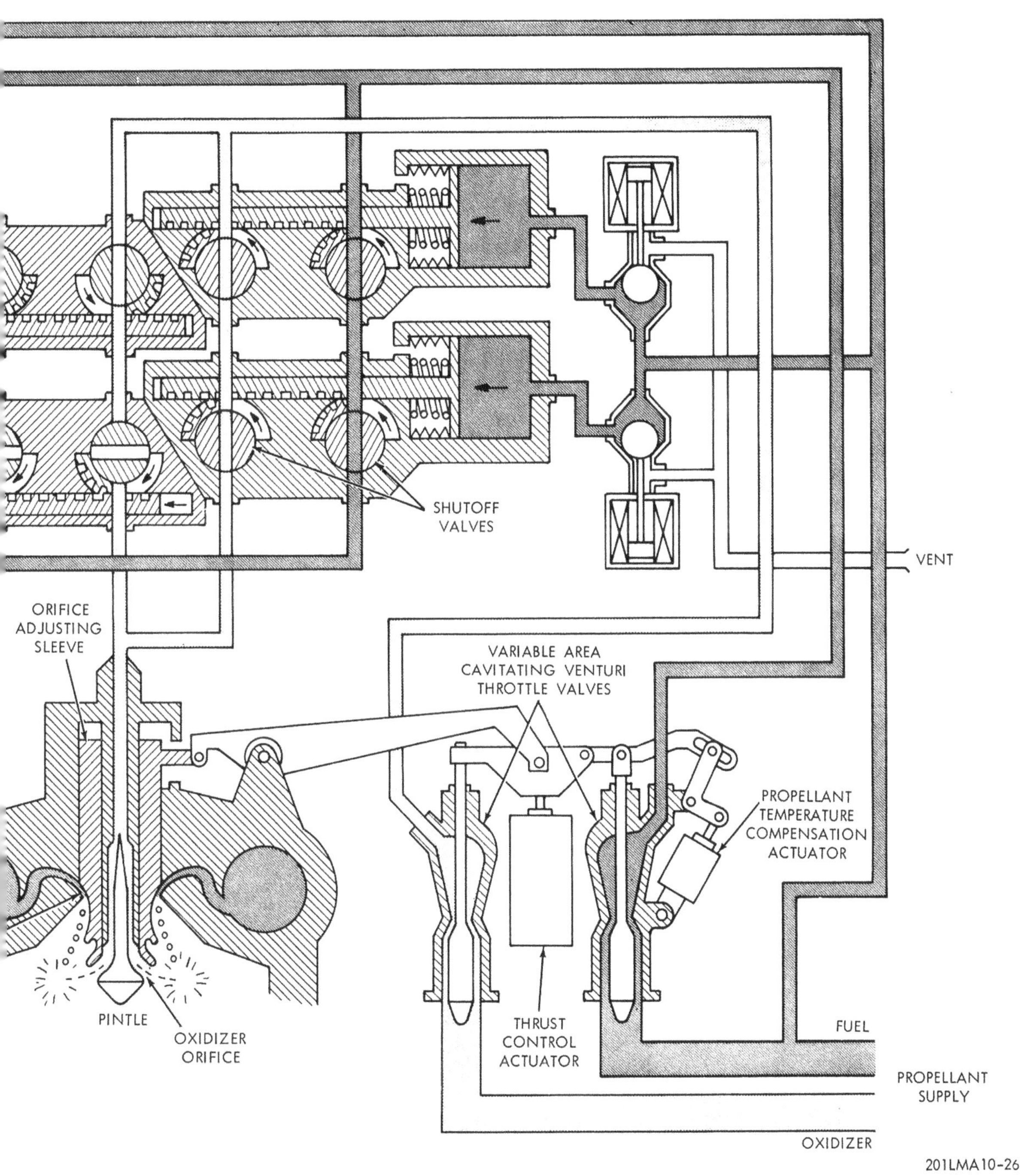

Figure 3-14. Propellant Injector and Valves (Mechanically Throttled Engine)

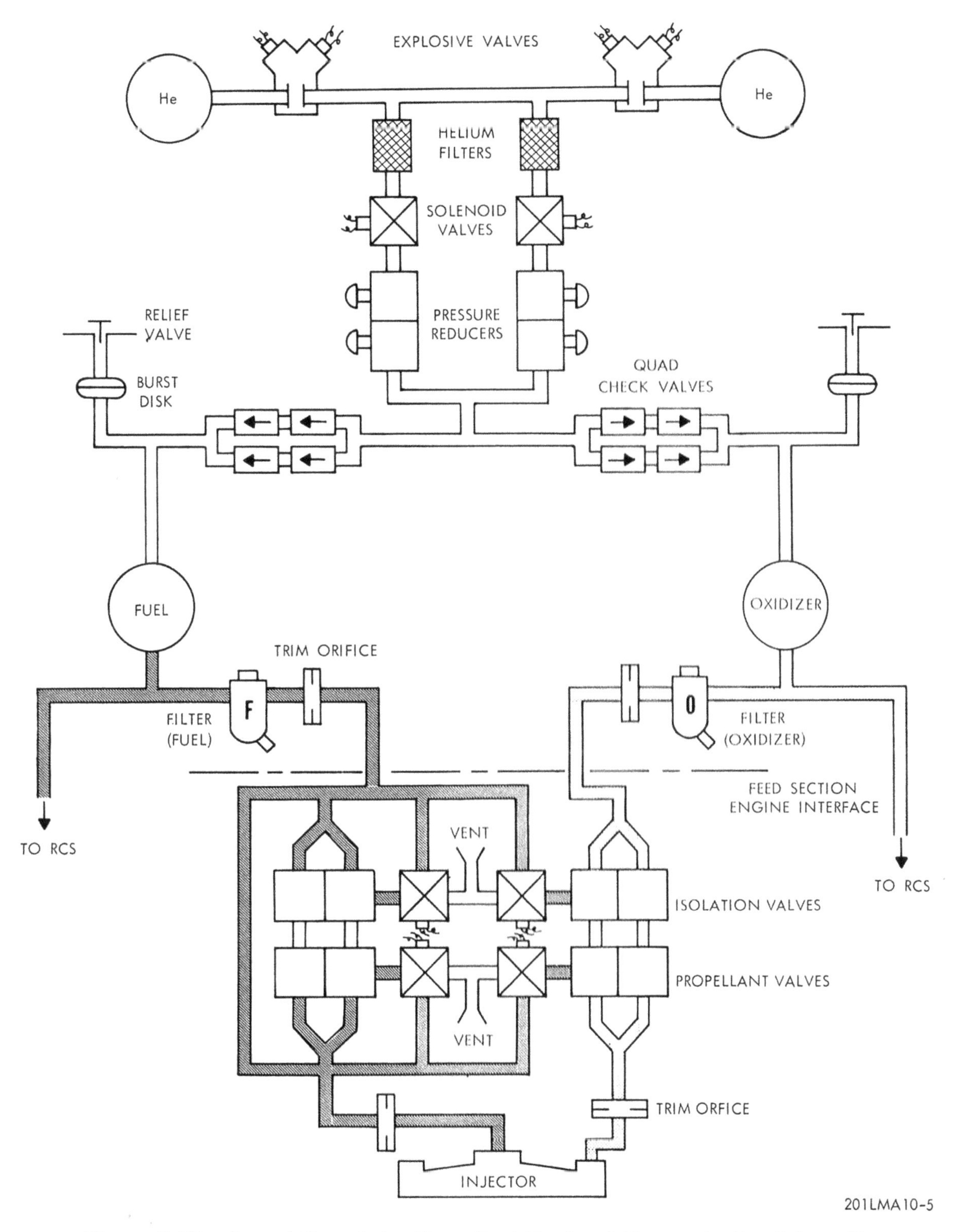

Figure 3-15. Ascent Propulsion Propellant Feed and Pressurization Schematic

closed, control is then obtained through the reducers in the line with the lower pressure setting. If an upstream regulator fails open, the downstream reducer continues to regulate the supply at its own pressure setting. Both are capable of handling full system load, however. A fail-open condition in a downstream regulator is negligible, since the upstream reducer is already in control. If both reducers in a line should fail open, the astronaut receives a propellant tank overpressurization indication, at which time he must close the solenoid valve in the malfunctioning line in order that normal pressure reduction can be obtained through the parallel line.

Downstream of the pressure reducer legs, the helium flow lines are manifolded together and then divide into two separate propellant tank pressurization lines, with a quadruple check valve in each. The check valves isolate the fuel and oxidizer tanks so that vapors from one tank cannot back up through the helium manifolds into the other tank prior to pressurization. A burst disk and relief valve is located in a branch in the helium pressure line to each tank for the purpose of preventing catastrophic tank overpressurization. The burst disk prevents possible helium loss, should the relief valve malfunction during normal pressurization. The helium is then piped into the baffled propellant tanks, where it acts directly on the surface of the fluids, forcing them through the system to the engine.

3-58. Ascent Engine. The ascent engine is a conventional, restartable, bipropellant rocket with an all-ablative nozzle extension, throat, and combustion chamber. Instrumentation includes sensors for measuring fuel and oxidizer inlet pressures and temperatures, injector inlet pressures, thrust chamber pressure, valve positions, vibration, and exterior surface temperatures.

Propellant flow to the engine combustion chamber is controlled through the valve package, trim orifice, and injector assemblies. At the feed section/engine interface, fuel and oxidizer lines divide and flow through a series-parallel valve arrangement containing ball valves in fuel/oxidizer pairs. Each pair is simultaneously opened and closed on a common crankshaft by an actuator that uses fuel as the actuating medium. Fuel flow is controlled through three-way solenoid valves. A schematic of the injector and valving arrangement is shown in figure 3-16.

At engine start, the solenoid valves are energized, lifting the caged balls from the fuel pressure ports and seating them against overboard vent ports. Fuel pressure then passes to the actuator chambers; the actuator pistons are extended, cranking the ball valves 90 degrees to the open position. Fuel and oxidizer then passes through trim orifices to adjust the thrust level and mixture ratio of the engine by trimming out the pressure drop of the propellant feed section, and is piped directly to the injector. The injector is a fixed-area, fixed-orifice design, with orifices arranged in a grid pattern on the injector face and drilled in triplets. Near the outer periphery of the face, orifices are drilled in doublets (one fuel and one oxidizer stream) to provide a temperature barrier near the ablative chamber wall.

At cutoff, the solenoids are deenergized, opening the actuator ports to the overboard vent. Residual fuel in the actuators is vented overboard into space and the actuators close under spring pressure, pulling the ball valves to the closed position.

3-59. INSTRUMENTATION SUBSYSTEM.

The Instrumentation Subsystem senses physical data, monitors the LEM subsystems during the unmanned and manned phases of the mission, performs an in-flight and lunar surface checkout, prepares LEM status data for transmission to earth,

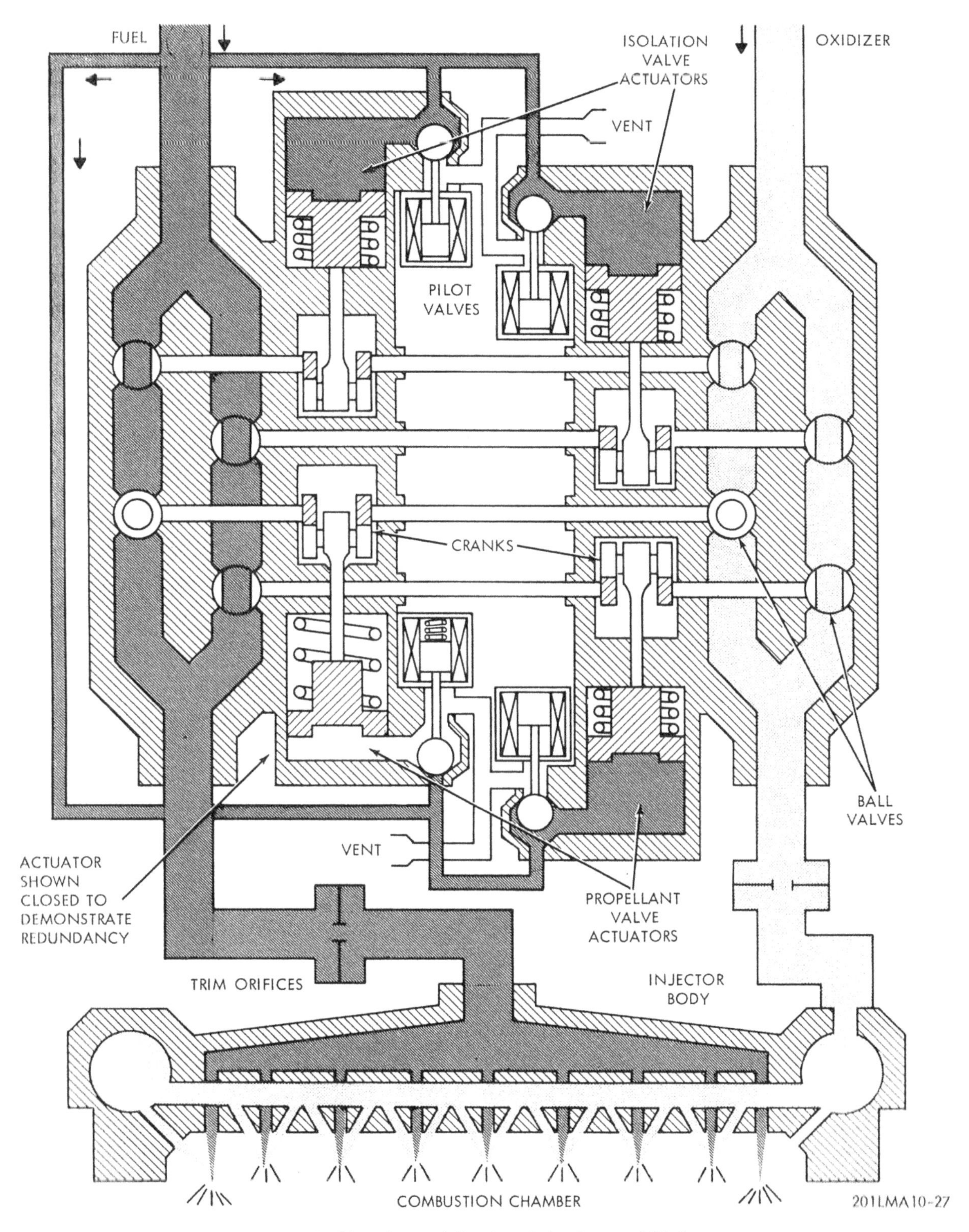

Figure 3-16. Ascent Engine Injector and Valves

provides timing frequencies for the LEM subsystems, and stores voice when the LEM is unable to transmit to earth. The instrumentation subsystem consists of sensors, the Signal Conditioning Electronics Assembly (SCEA), Caution and Warning Electronics Assembly (CWEA), On-Board Checkout Electronics Assembly (OBCEA), Pulse Code Modulation and Timing Electronics Assembly (PCMTEA), and Data Storage Electronics Assembly (DSEA). The subsystem equipment operates on 28 volts dc and 115-volt, 400 cps, single-phase power supplied by the Electrical Power Subsystem. The instrumentation subsystem provides the crew and ground facilities with LEM performance data during all phases of the mission. This enhances crew safety and mission success.

The operation of the instrumentation subsystem is discussed on an equipment basis in the following paragraphs. The assemblies that comprises the instrumentation subsystem are shown as solid lined blocks in figure 3-17.

3-60. SENSORS AND SIGNAL CONDITIONING ELECTRONICS ASSEMBLY.

The sensors and SCEA provide the capability for gathering the various physical data in the LEM subsystems and modifying them into a form compatible with the instrumentation subsystem equipment. The sensors are permanently installed in the various LEM subsystems. They sense the various physical data (such as temperature, valve action, pressure, switch positions, voltages, currents) and convert the data into signals and present them to the SCEA. The SCEA conditions the signals into a form compatible with its output equipment. This equipment contains the various types of signal modifiers, d-c amplifiers, d-c attenuators, a-c to d-c converters, etc., to condition the variety of input signals. In addition to the signals from the sensors, the SCEA receives signals directly from other components within the LEM. Not all signals presented to the SCEA require conditioning. Signals that do not require conditioning, together with those requiring conditioning, are fed to one or more of the following: OBCEA, CWEA, and PCMTEA.

3-61. CAUTION AND WARNING ELECTRONICS ASSEMBLY.

The CWEA provides the crew with a rapid check of LEM status during the manned phase of the mission. The input data from the SCEA is continuously monitored by this equipment, to detect a malfunction. If a malfunction is detected, the CWEA provides signals to the Control and Display Subsystem and actuates the audio tone generator, lights the caution or warning indicators, and lights the indicator associated with the malfunctioning equipment. These three indications alert the crew and aids them in isolating the malfunction. There are two types of malfunction indications: one is a warning; the other, a caution. A warning indication means that a malfunction that affects crew safety and requires immediate action has occurred. A caution indication means that a malfunction that does not require immediate action has occurred. Signals that reflect malfunctions are applied to the PCMTEA to be telemetered to earth.

3-62. ON-BOARD CHECKOUT ELECTRONICS ASSEMBLY.

The OBCEA provides the crew with a central checkout facility for in-flight and lunar surface checkout of the LEM subsystems. This equipment contains comparators, stored limits, and logic circuitry for testing and analyzing the LEM subsystems. The OBCEA isolates a malfunction to either the equipment, assembly, or subassembly, and provides visual display of subsystem status during checkout. Checkout of the

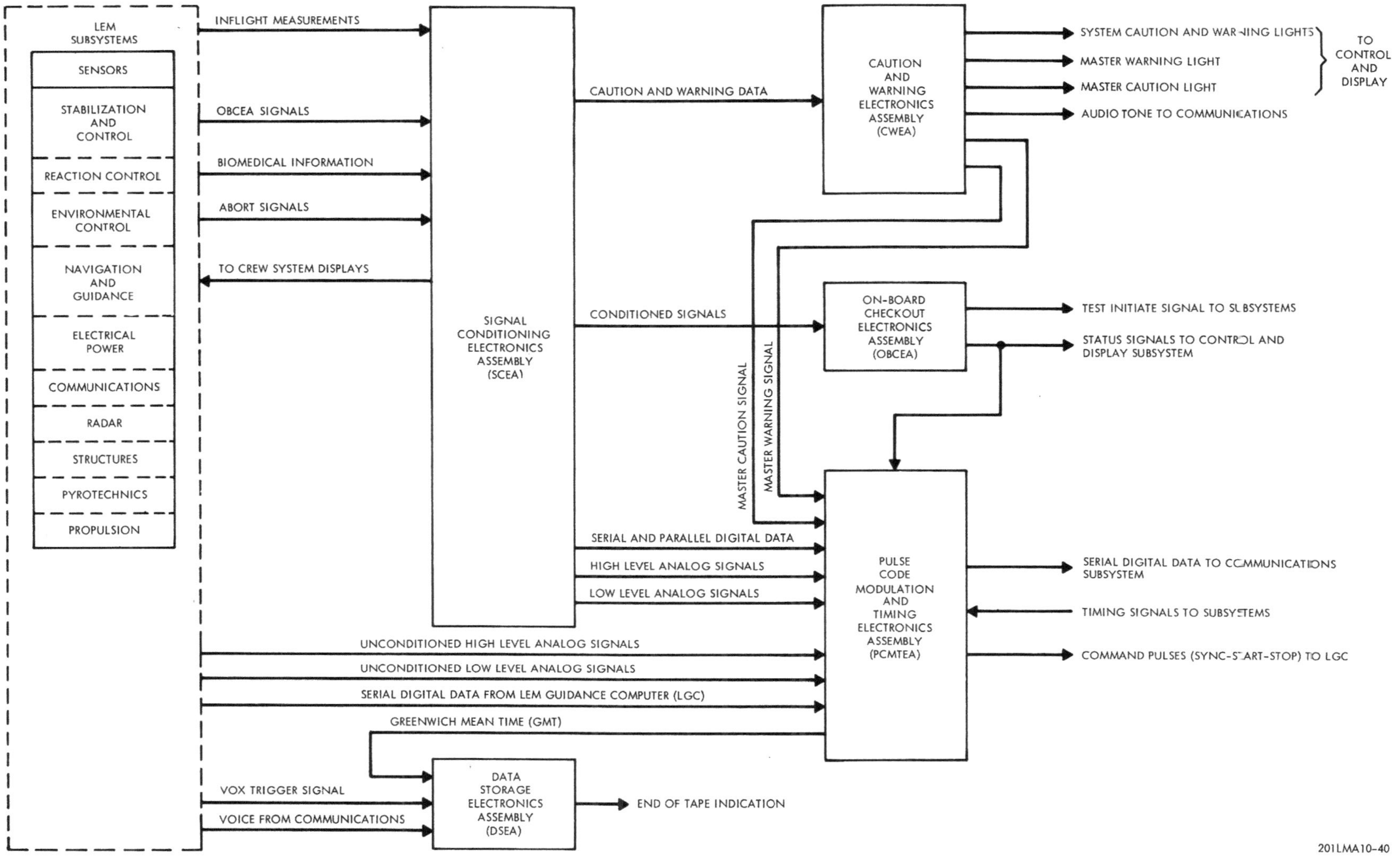

LEM SUBSYSTEMS
SENSORS
STABILIZATION AND CONTROL
REACTION CONTROL
ENVIRONMENTAL CONTROL
NAVIGATION AND GUIDANCE
ELECTRICAL POWER
COMMUNICATIONS
RADAR
STRUCTURES
PYROTECHNICS
PROPULSION
INFLIGHT MEASUREMENTS
OBCEA SIGNALS
BIOMEDICAL INFORMATION
ABORT SIGNALS
TO CREW SYSTEM DISPLAYS
SIGNAL CONDITIONING ELECTRONICS ASSEMBLY (SCEA)
CAUTION AND WARNING DATA
CONDITIONED SIGNALS
SERIAL AND PARALLEL DIGITAL DATA
HIGH LEVEL ANALOG SIGNALS
LOW LEVEL ANALOG SIGNALS
MASTER CAUTION SIGNAL
MASTER WARNING SIGNAL
CAUTION AND WARNING ELECTRONICS ASSEMBLY (CWEA)
SYSTEM CAUTION AND WARNING LIGHTS
MASTER WARNING LIGHT
MASTER CAUTION LIGHT
TO CONTROL AND DISPLAY
AUDIO TONE TO COMMUNICATIONS
ON-BOARD CHECKOUT ELECTRONICS ASSEMBLY (OBCEA)
TEST INITIATE SIGNAL TO SUBSYSTEMS
STATUS SIGNALS TO CONTROL AND DISPLAY SUBSYSTEM
PULSE CODE MODULATION AND TIMING ELECTRONICS ASSEMBLY (PCMTEA)
SERIAL DIGITAL DATA TO COMMUNICATIONS SUBSYSTEM
TIMING SIGNALS TO SUBSYSTEMS
COMMAND PULSES (SYNC-START-STOP) TO LGC
UNCONDITIONED HIGH LEVEL ANALOG SIGNALS
UNCONDITIONED LOW LEVEL ANALOG SIGNALS
SERIAL DIGITAL DATA FROM LEM GUIDANCE COMPUTER (LGC)
GREENWICH MEAN TIME (GMT)
VOX TRIGGER SIGNAL
VOICE FROM COMMUNICATIONS
DATA STORAGE ELECTRONICS ASSEMBLY (DSEA)
END OF TAPE INDICATION
201LMA10-40

Figure 3-17. Instrumentation Subsystem Block Diagram

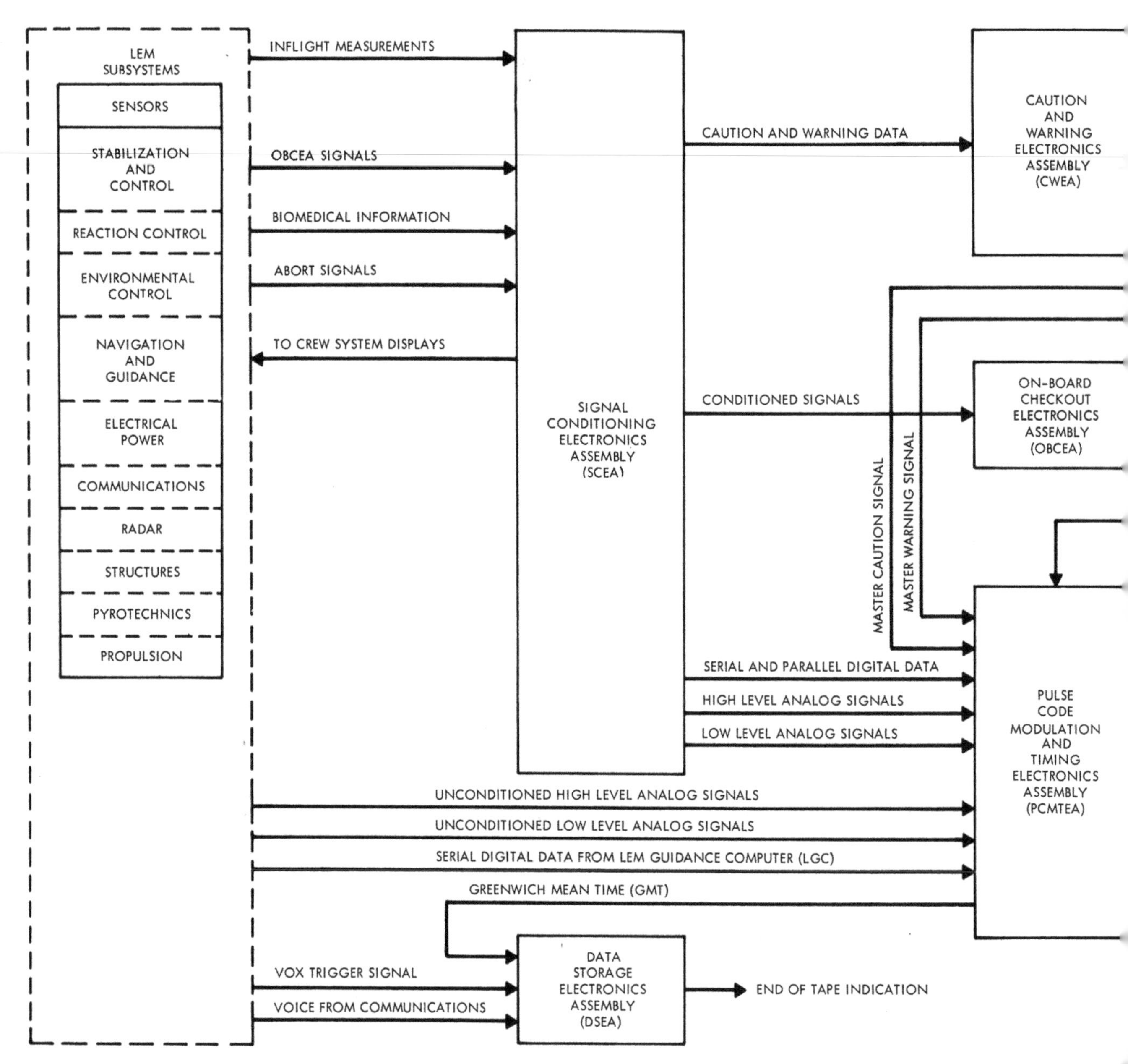
LEM SUBSYSTEMS
SENSORS
STABILIZATION AND CONTROL
REACTION CONTROL
ENVIRONMENTAL CONTROL
NAVIGATION AND GUIDANCE
ELECTRICAL POWER
COMMUNICATIONS
RADAR
STRUCTURES
PYROTECHNICS
PROPULSION
INFLIGHT MEASUREMENTS
OBCEA SIGNALS
BIOMEDICAL INFORMATION
ABORT SIGNALS
TO CREW SYSTEM DISPLAYS
SIGNAL CONDITIONING ELECTRONICS ASSEMBLY (SCEA)
CAUTION AND WARNING DATA
CAUTION AND WARNING ELECTRONICS ASSEMBLY (CWEA)
CONDITIONED SIGNALS
ON-BOARD CHECKOUT ELECTRONICS ASSEMBLY (OBCEA)
MASTER CAUTION SIGNAL
MASTER WARNING SIGNAL
SERIAL AND PARALLEL DIGITAL DATA
HIGH LEVEL ANALOG SIGNALS
LOW LEVEL ANALOG SIGNALS
PULSE CODE MODULATION AND TIMING ELECTRONICS ASSEMBLY (PCMTEA)
UNCONDITIONED HIGH LEVEL ANALOG SIGNALS
UNCONDITIONED LOW LEVEL ANALOG SIGNALS
SERIAL DIGITAL DATA FROM LEM GUIDANCE COMPUTER (LGC)
GREENWICH MEAN TIME (GMT)
VOX TRIGGER SIGNAL
VOICE FROM COMMUNICATIONS
DATA STORAGE ELECTRONICS ASSEMBLY (DSEA)
END OF TAPE INDICATION

Figure 3-1

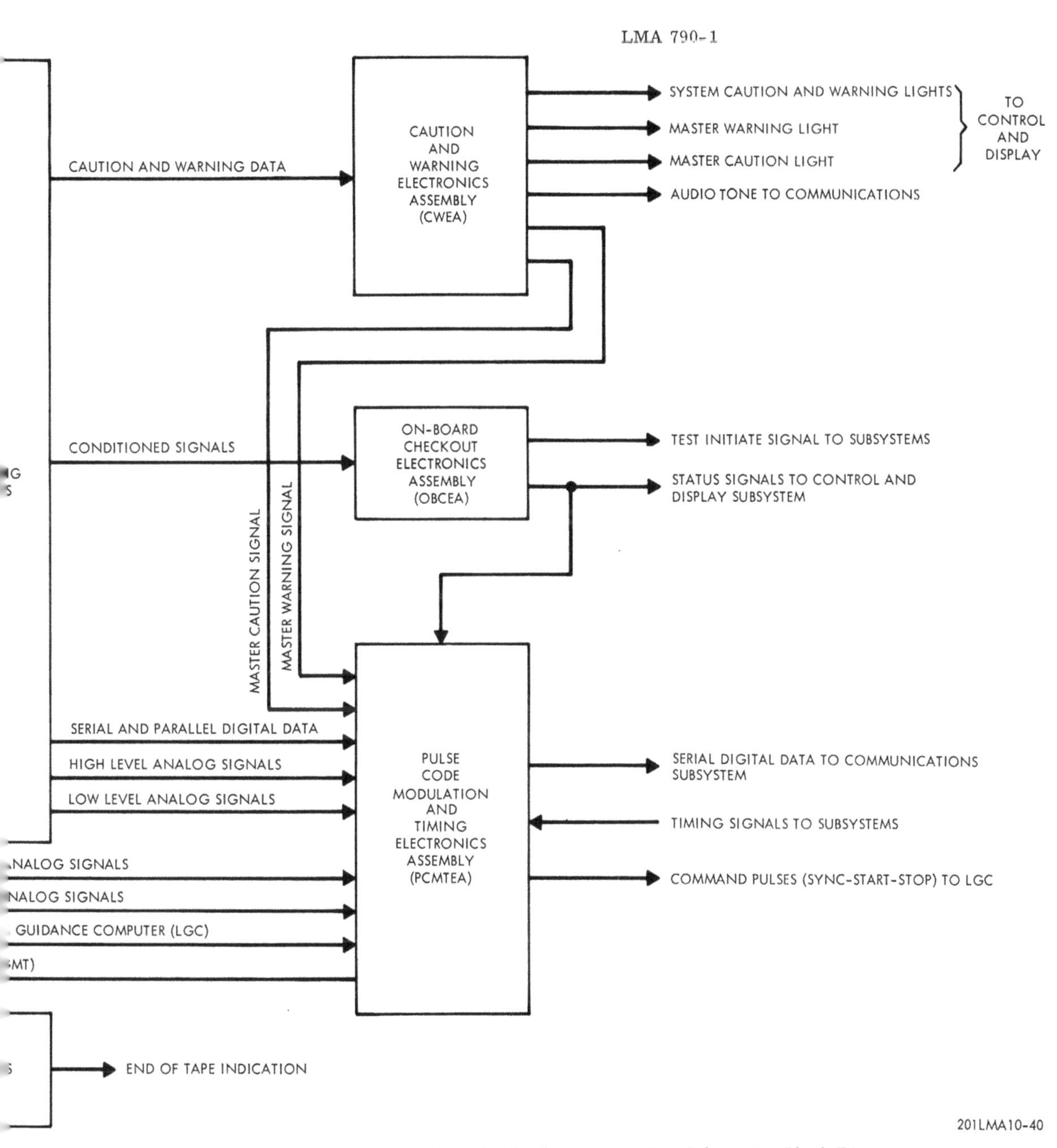

Figure 3-17. Instrumentation Subsystem Block Diagram

LEM subsystems are performed during post transposition, lunar orbit, post lunar landing, and lunar prelaunch. Post transposition checkout is performed to ensure that all subsystems withstood the launch, translunar injection, and transposition. Lunar orbit checkout is performed to ensure that all subsystems are functioning properly before separation and descent to the lunar surface. Post lunar landing checkout is performed to ensure that all subsystems are functioning properly before ascent. Signals that reflect all checkout malfunctions are applied to the PCMTEA to be telemetered to earth.

3-63. PULSE CODE MODULATION AND TIMING ELECTRONICS ASSEMBLY.

The PCMTEA changes all LEM data into signals for transmission to earth and provides a time reference for any stored data. This equipment contains: analog multiplexers, amplifiers, analog-to-digital converters, coders, digital multiplexers, digital registers, a programmer, and a timing generator. It combines analog inputs and parallel and serial digital inputs into a serial digital data train for presentation to earth, and to the automatic checkout equipment (ACE). This data can be presented at one of two selectable rates of 1,600 bits per second or 51,200 bits per second. The timing generator is used to provide timing frequencies for guidance, telemetry, calibration, and displays.

3-64. DATA STORAGE ELECTRONICS ASSEMBLY.

The DSEA provides tape storage for voice communications and time correlation. The DSEA is used as an automatic "notepad" by the astronauts. Voice signals, from the astronaut's microphones, are processed in the Audio Center Assembly (ACA) and fed to the DSEA which is activated by a voice-operated circuit (VOX) trigger signal. The DSEA has no playback capability. All recorded tapes will be carried back to the Command/Service module by the astronauts.

3-65. COMMUNICATIONS SUBSYSTEM. (See figure 3 -18).

The Communications Subsystem provides the link between the lunar astronauts, the orbiting Command/Service/modules (C/SM), and earth monitoring stations. The subsystem contains two radio frequency (RF) sections, one operating in the VHF range and the other in the UHF range (S-Band), a television section, and a signal processing section.

The RF sections provide two-way voice communications between the LEM and the C/SM, LEM and earth, LEM and the extravehicular astronaut, and between crewmembers within the LEM. In addition, these sections receive and retransmit tracking and ranging information from earth and transmit, to earth, television, telemetry, and biomedical information, and emergency code keying in the advent of voice transmission failure.

The television section is utilized by the extravehicular astronaut to televise the lunar surface within an eighty-foot radius, approximately, of the grounded vehicle.

The signal processing section modulates, switches, amplifies, and multiplexes the various signals and inputs from other subsystems during the different operating modes.

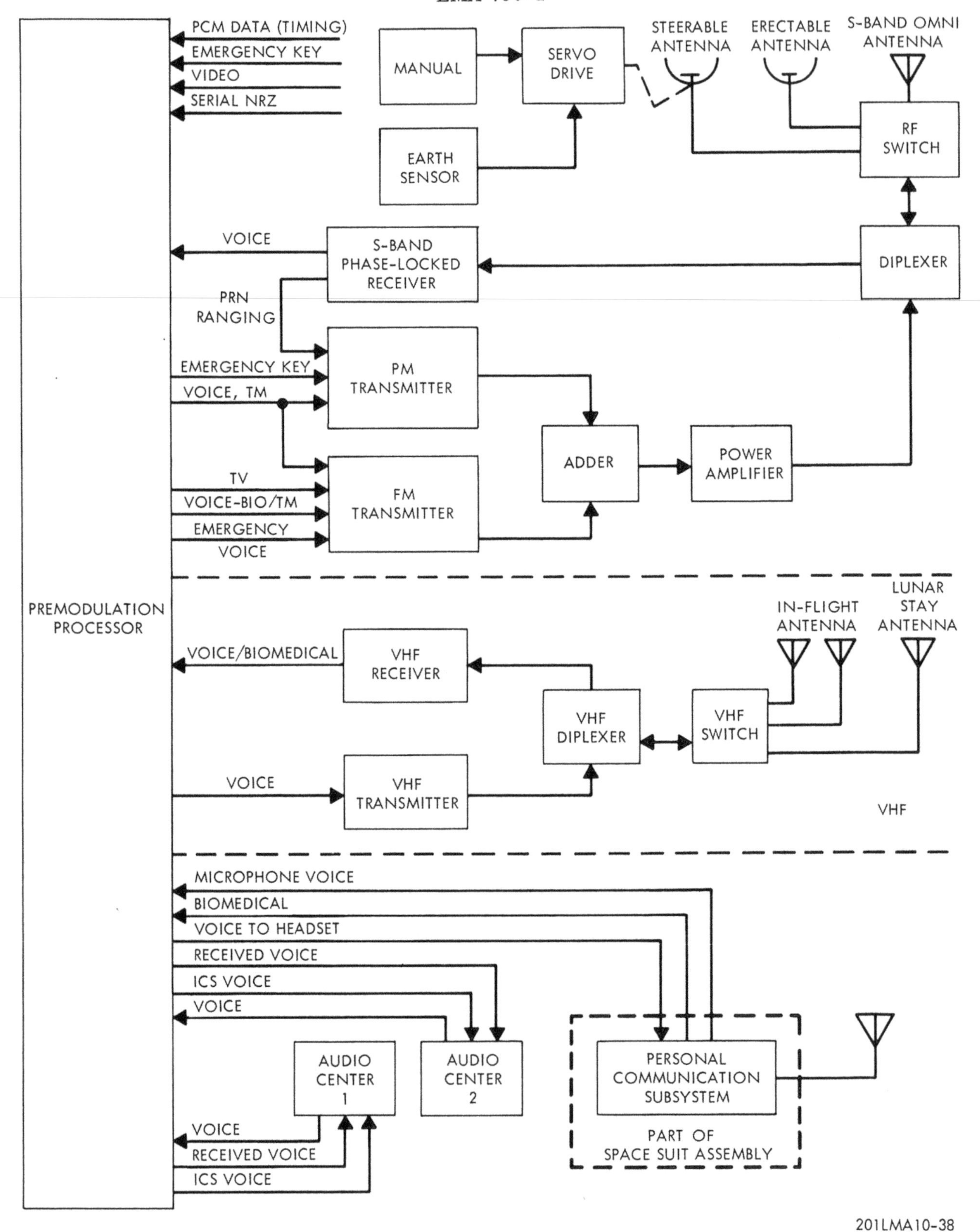

Figure 3-18. Communications Subsystem Block Diagram

3-66. S-BAND SECTION.

The S-band section provides the LEM-earth communication link. The S-band equipment comprises a transceiver, a power amplifier, a diplexer, an RF switch, a steerable antenna with appropriate control and servo circuits, an erectable antenna, and an emergency omnidirectional antenna. Television, voice, telemetry, biomedical, emergency keying, and tracking and ranging signals are accommodated by this equipment.

3-67. VHF SECTION.

The VHF equipment, which includes a VHF/AM transceiver, a VHF diplexer, an RF switch, an in-flight antenna, and a lunar-stay antenna, provides the LEM-man and LEM-C/SM communication links.

3-68. SIGNAL PROCESSING SECTION.

The signal processing section consists of the premodulation processor (PMP) and two audio centers. The section receives signals from the Instrumentation Subsystem and the VHF and S-band sections and processes them for transmission.

3-69. S-BAND COMMUNICATIONS.

The S-band communications section receives information from the PMP, in some cases directly, and in other cases in response to control settings made by the astronaut. This information is applied to the inputs of the S-band transceiver, which is a solid-state equipment consisting of a phase-modulated (PM) transmitter, a frequency-modulated (FM) transmitter, and a phase-locked receiver. The information is transmitted from one of the transmitter assemblies of the transceiver to the power amplifier, which provides an RF power amplification factor of approximately 27. The information is then conducted to the S-band diplexer, which permits simultaneous reception and transmission from a single antenna, and then through the RF switch to one of the three S-band antennas. When less gain is required, the power amplifier is bypassed and the signals routed directly to the antenna.

Signal reception from earth is confined to voice and psuedo-random noise (PRN) ranging signals. The signal flow is from the antenna through the RF switch, diplexer, phase-locked receiver, premodulation processor, audio center, to the astronaut's headset or through the VHF transmitter to the roving astronaut on the lunar surface. All S-band communication is accomplished during line-of-sight phases of the mission.

3-70. Steerable Antenna. The steerable antenna is a medium gain, unidirectional antenna that provides hemispherical coverage about its mounting axis. It is mounted to the LEM structure and is used during the LEM lunar orbit, descent, ascent, rendezvous, and docking phases of the mission. The antenna is mounted on a double elevation gimbal that is servo-controlled in response to automatic or manual slewing signals. Switches on the control panel provide mode selection and slewing commands to the antenna servos. An infrared earth sensor, and an RF signal strength meter provide tracking error signals in the automatic mode.

3-71. Erectable Antenna. The erectable antenna is a high-gain unit comprising a helix-fed parabaloidal reflector mounted on a tripod support and an aiming telescope.

This antenna is used during the lunar stay to transmit and receive S-band communications between LEM and earth.

3-72. VHF COMMUNICATIONS.

The VHF communications link provides two-way voice communications between the LEM and the C/SM, and the LEM and the extravehicular astronaut. Voice transmissions are processed through the audio centers and are coupled to the VHF/AM transceiver. From the transceiver, the signal is coupled to the in-flight antenna or lunar-stay antenna through the VHF diplexer and RF switch. Voice transmissions from the C/SM are coupled from the appropriate antenna to the receiver portion of the transceiver through the RF switch and the diplexer. The receiver output is processed in the PMP and the audio centers and is fed to the astronauts' headsets or routed back to the audio centers for VHF transmission to the extravehicular astronaut. All VHF communication is accomplished during line-of-sight phases of the mission.

Communications between the LEM and the extravehicular astronaut are accomplished through the VHF section. The VHF transceiver receives voice and biomedical information concerning the physiological functions of the astronaut and the operation of his PLSS while he is conducting lunar explorations. Voice information is monitored by the astronaut in the LEM and voice and biomedical information transmitted via the S-band link to earth monitors. A television camera is connected to the vehicle through an umbilical and can be used outside the LEM.

3-73. Intercommunications. A closed-circuit intercommunications system permits voice communications when both astronauts are aboard the LEM.

3-74. Personal Communication System. The Personal Communication System is part of the astronaut's space suit. It provides the dual capability of transmitting and receiving voice only (simplex operation, emergency mode) or receiving voice and transmitting voice, biomedical, and suit data, (duplex operation, normal mode).

The system contains two transmitters and two receivers, one set for each mode of operation, and other components (mixer, amplifier, diplexer, etc.) associated with duplex operation. The output of the system is transmitted via the suit antenna.

3-75. SIGNAL PROCESSING SECTION.

The signal processing equipment, consisting of the premodulation processor and the two audio centers, provides the interface components for processing signals to be transferred between the various data gathering equipment external to and within the LEM and the various receivers and transmitters. The processing functions include multiplexing, filtering, amplifying, and switching of voice, telemetry, and television signals.

3-76. Premodulation Processor. The PMP contains filters, subcarrier oscillators, matching networks, the voltage controlled frequency generator and modulator, and switches. It performs signal modulation and mixing and accomplishes signal switching so that the correct intelligence corresponding to a given mode of operation is transmitted.

3-77. Audio Centers. The audio centers are two identical assemblies. These units operate with complementary panel controls to provide selection, isolation, and amplification of LEM audio to and from the subsystem transceivers.

3-78. CONTROLS AND DISPLAYS.

All controls and displays for the Communications Subsystem are on the Left and Right Instrument Panels in the crew compartment. The panels are illustrated in figure 3-2.

3-79. ELECTRICAL POWER SUBSYSTEM.

The Electrical Power Subsystem (EPS) (see figure 3-19) provides power to all circuits in the LEM. Power originates in three fuel cell assemblies, or, under emergency conditions, in an auxiliary battery. The three fuel cell assemblies can act independently, or in parallel, to provide 28-volt dc to the monitor bus and to the essential buses.

In parallel operation, the three fuel cell assemblies operate with maximum efficiency to provide power to the three buses. Under nonparallel operating conditions, each fuel cell assembly energizes an essential or the monitor bus. If any fuel cell assembly fails, a logic network in the feeder control assembly automatically deenergizes the monitor bus so that the operating fuel cell assembly provides power only to the two essential buses. With one fuel cell assembly operating, the monitor bus may be manually reenergized by a manually operated override switch. The monitor bus override switch is closed only after all absolutely nonessential loads have been removed from the circuit, to prevent overloading the single operating fuel cell assembly.

If all fuel cell assemblies fail, the auxiliary battery automatically energizes the essential bus, which distributes power to all critical survival and pyrotechnic circuits. The kilowatt-hour capacity of the silver-zinc auxiliary battery is adequate for all survival loads during lunar launch, orbital rendezvous, and docking phases of the mission.

The monitor bus, in addition to providing power to nonessential d-c loads, transmits dc to the inverter, where it is transformed into 115/200-volt, three-phase, 400-cps ac. The a-c power passes through the a-c bus, where it is distributed to all a-c loads through a four-wire, neutral grounded system.

3-80. FUEL CELL ASSEMBLY.

Each fuel cell assembly (see figure 3-20) consists of two subassemblies: the reactor subassembly and the fuel cell control subassembly. The reactor subassembly contains 30 individual fuel cells, cell and stack heaters, and a manifold system for the delivery of hydrogen, oxygen, and nitrogen, for the purging of oxygen and nitrogen, and for venting hydrogen. The 30 fuel cells are arranged in two stacks of 15. Each individual fuel cell contains two porous nickel-nickel oxide electrodes, aqueous potassium hydroxide electrolyte, a heating unit, reactant and inert gas passages, and all necessary fittings for manifolding to other cells. The liquid hydrogen and oxygen reactants are cryogenically stored in tanks. When the LEM fuel cell assembly is in operation, the reactants ionize, release electrons, and form water. To remove the water and heat produced during this process, an excess of hydrogen is passed through the hydrogen gas space and vented overboard.

The electrodes are of dual-porosity nickel-nickel oxide and provide a large interface with which reactant and liquid electrolyte is in contact. Each electrode is made with a thin, fine-pore layer that is in contact with the electrolyte, and a relatively thick, coarse-pore layer that is in contact with a reactant. The configuration of the electrodes is such that mixing of hydrogen and oxygen gas will not occur as long as the electrolyte is in the liquid state.

To speed up the reaction process and to keep the potassium hydroxide electrolyte liquified, the fuel cells are operated at a temperature of 260°C (500°F). This temperature is maintained by two sets of heaters in the cell stack assembly. In addition, the cell heater separates the hydrogen electrode structure of one cell from the hydrogen electrode structure of the adjacent cell.

The manifold system in the reactor subassembly provides six manifolds per stack of cells. These are the hydrogen, oxygen, and nitrogen delivery manifolds, oxygen and nitrogen purge manifolds, and a hydrogen vent manifold. Each fuel cell has an inert blanket of nitrogen around its electrodes and its gas reactant passages.

3-81. CRYOGENIC HYDROGEN AND OXYGEN STORAGE AND SUPPLY.

Cryogenic storage and supply tanks contain the hydrogen and oxygen necessary for operation of the fuel cells of the EPS and for the oxygen consumption requirements of the Environmental Control Subsystem. A hydrogen tank assembly and an oxygen tank assembly supply flow to the fuel cell assembly during all phases of the LEM mission. The hydrogen tank (TH-2) and the oxygen tank (TO-2) in the LEM descent stage contain sufficient reactants to power the fuel cell assemblies through the lunar descent maneuver and during lunar stay time. A second hydrogen tank (TH-1) and a second oxygen tank (TO-1) are located in the ascent stage and are capable of satisfying basic mission electrical power requirements from lunar ascent through orbital docking. If the mission is aborted during the lunar descent maneuver, the descent stage is ejected and the TH-1 and TO-1 tanks provide reactant to the fuel cell assemblies for the rendezvous and docking maneuvers.

Supercritical hydrogen and supercritical oxygen is pressurized by electrical heating elements in both the descent stage and ascent stage storage tanks. The reactants are heated by passing through the ECS heat exchanger and then passing into the fuel cell assemblies. A series of latch-type solenoid valves control the flow.

3-82. FUEL CELL CONTROL SUBASSEMBLY.

The control subassembly contains the various components, controls, and protective devices that control the operation of the fuel cell assembly. It is mounted beneath the reactor subassembly and is shielded from heat transfer by a layer of insulation. There is an isolation inlet valve for each reactant to prevent contamination of the hydrogen and oxygen lines during prelaunch storage and to isolate the LEM hydrogen and oxygen supply from the fuel cell assembly if a malfunction occurs.

3-83. ELECTRICAL CONTROLS.

The fuel cell assembly electrical controls consist of the following units: control relay, reverse current relay, overcurrent relay, overtemperature switch, pressure and temperature sensors, emergency shutdown switch, and an interlock system.

The control relay receives a signal when a reverse current, overcurrent, or overtemperature situation exists while the fuel cell assembly is operating in the automatic mode. It also receives a signal from the emergency shutdown switch when the switch is actuated. Any of these four signals causes the control relay to isolate and shut down the fuel cell assembly. The reverse current relay is in series with the power lead of the fuel cell assembly and contains a series coil, sensitive to reverse current flow. This coil, when energized, actuates an armature to close two pairs of normally open contacts. One pair sends a signal to the control relay

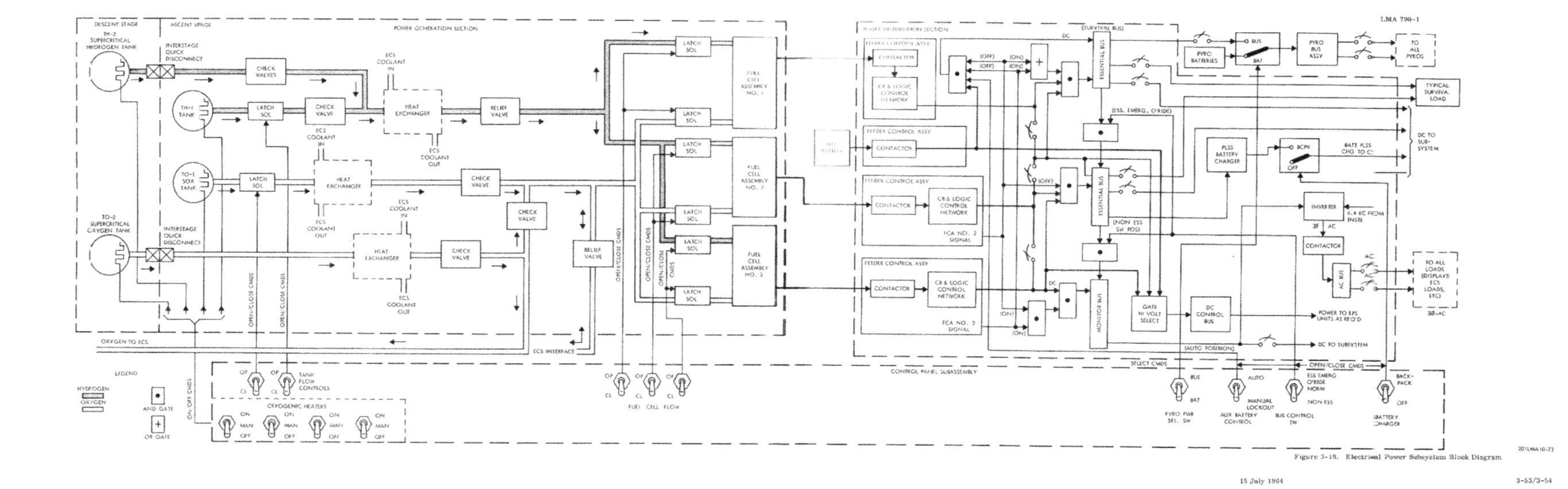

Figure 3-19. Electrical Power Subsystem Block Diagram

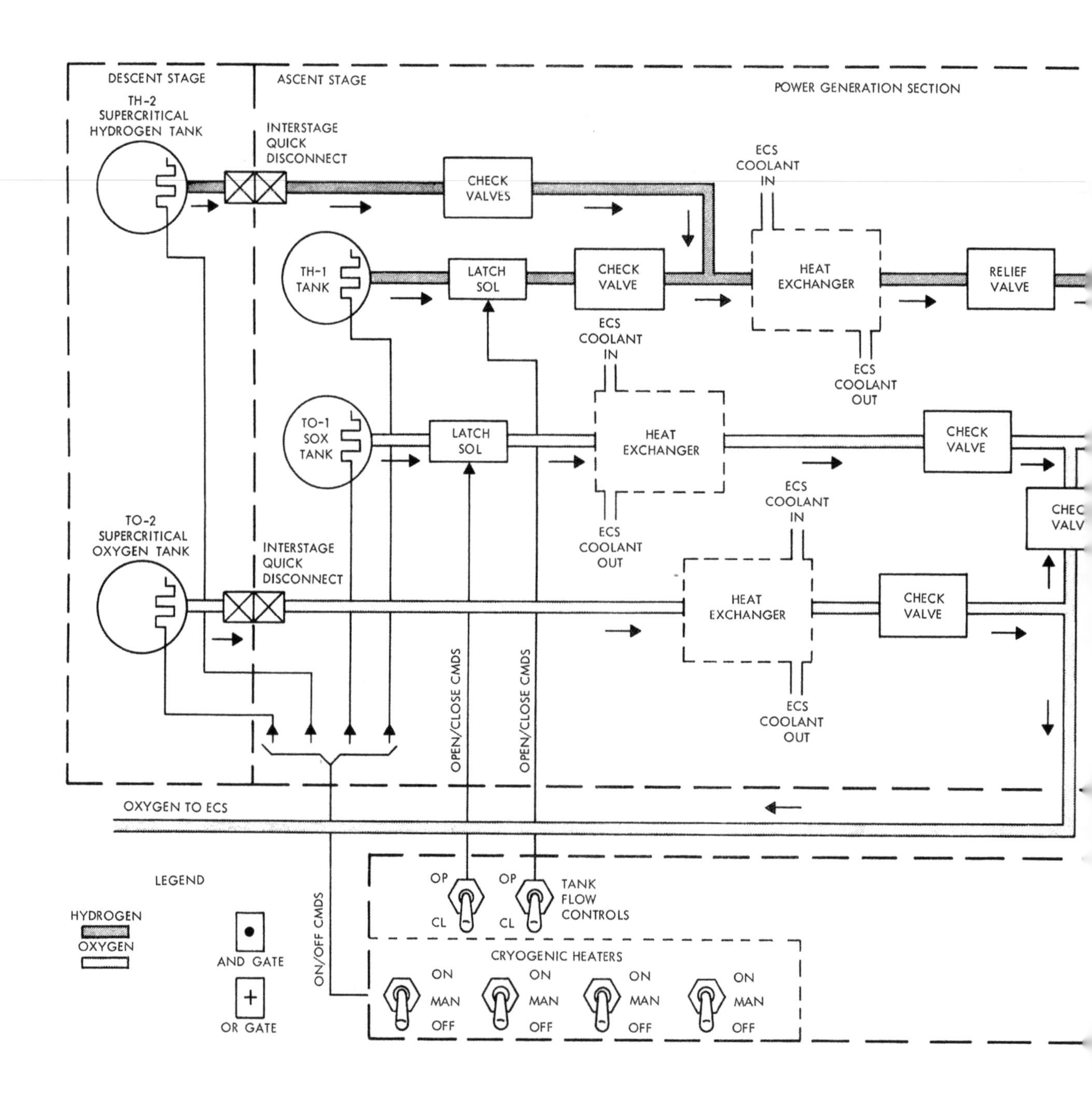

DESCENT STAGE
ASCENT STAGE
POWER GENERATION SECTION
TH-2
SUPERCRITICAL
HYDROGEN TANK
INTERSTAGE
QUICK
DISCONNECT
CHECK
VALVES
ECS
COOLANT
IN
TH-1
TANK
LATCH
SOL
CHECK
VALVE
HEAT
EXCHANGER
RELIEF
VALVE
ECS
COOLANT
IN
ECS
COOLANT
OUT
TO-1
SOX
TANK
LATCH
SOL
HEAT
EXCHANGER
CHECK
VALVE
ECS
COOLANT
IN
ECS
COOLANT
OUT
TO-2
SUPERCRITICAL
OXYGEN TANK
INTERSTAGE
QUICK
DISCONNECT
HEAT
EXCHANGER
CHECK
VALVE
ECS
COOLANT
OUT
OPEN/CLOSE CMDS
OPEN/CLOSE CMDS
OXYGEN TO ECS
LEGEND
HYDROGEN
OXYGEN
AND GATE
OR GATE
ON/OFF CMDS
OP
CL
OP
CL
TANK
FLOW
CONTROLS
CRYOGENIC HEATERS
ON
MAN
OFF
ON
MAN
OFF
ON
MAN
OFF
ON
MAN
OFF

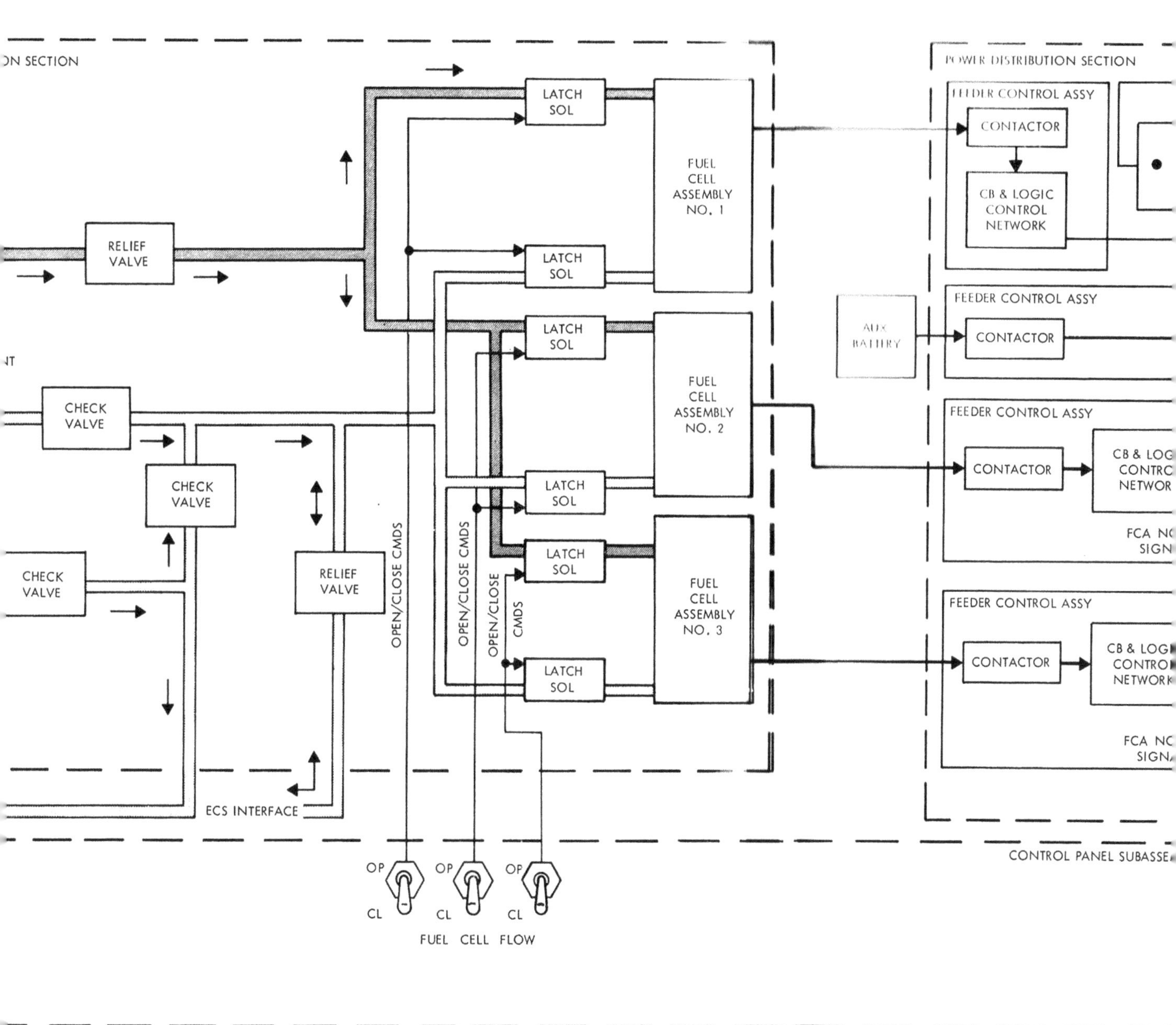

ON SECTION
RELIEF VALVE
LATCH SOL
LATCH SOL
LATCH SOL
LATCH SOL
LATCH SOL
LATCH SOL
FUEL CELL ASSEMBLY NO. 1
FUEL CELL ASSEMBLY NO. 2
FUEL CELL ASSEMBLY NO. 3
CHECK VALVE
CHECK VALVE
CHECK VALVE
RELIEF VALVE
OPEN/CLOSE CMDS
OPEN/CLOSE CMDS
OPEN/CLOSE CMDS
ECS INTERFACE
POWER DISTRIBUTION SECTION
FEEDER CONTROL ASSY
CONTACTOR
CB & LOGIC CONTROL NETWORK
FEEDER CONTROL ASSY
AUX BATTERY
CONTACTOR
FEEDER CONTROL ASSY
CONTACTOR
FEEDER CONTROL ASSY
CONTACTOR
CONTROL PANEL SUBASSE
OP
CL
OP
CL
OP
CL
FUEL CELL FLOW

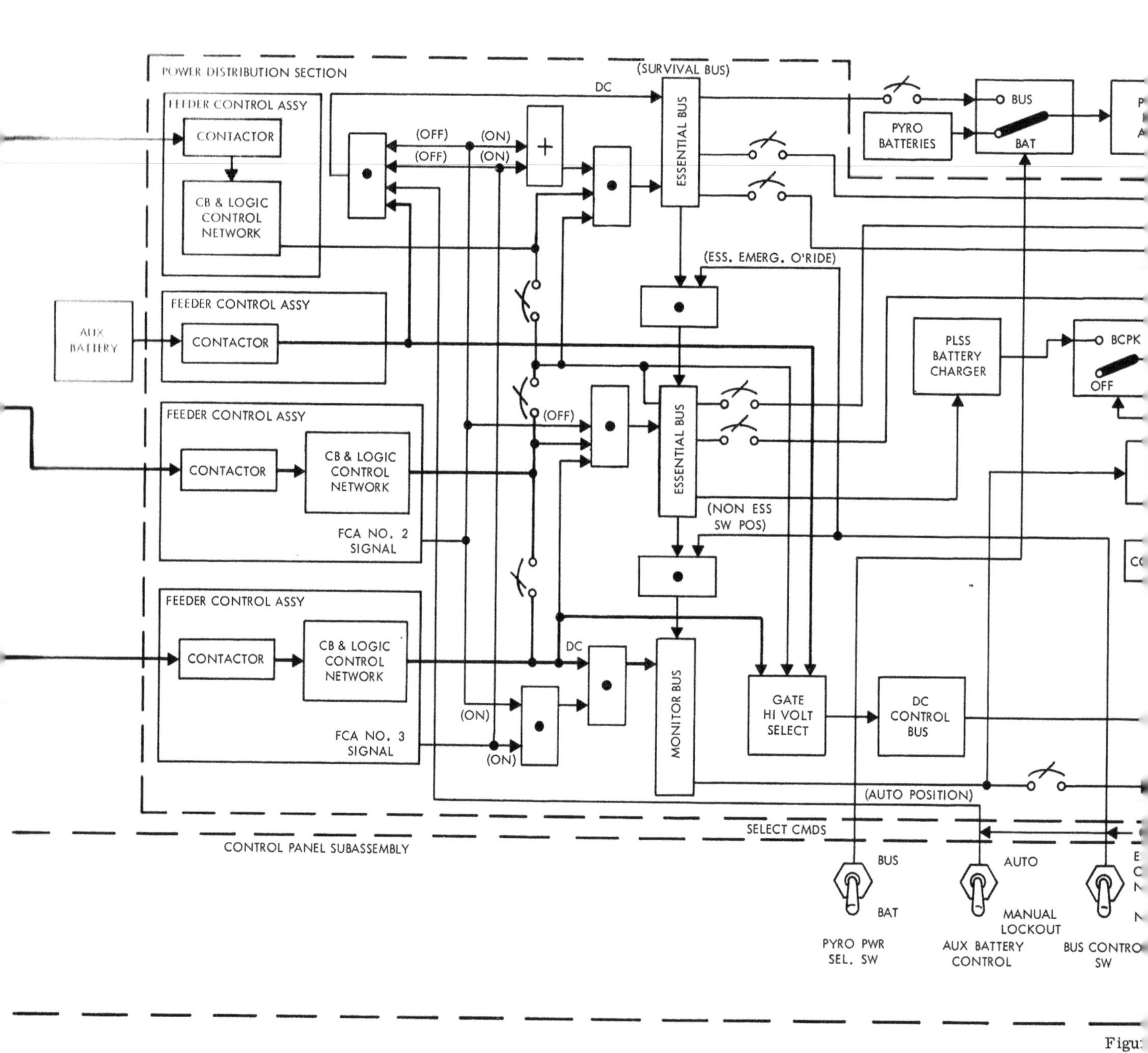

POWER DISTRIBUTION SECTION
(SURVIVAL BUS)
DC
FEEDER CONTROL ASSY
CONTACTOR
CB & LOGIC CONTROL NETWORK
(OFF)
(ON)
(OFF)
(ON)
ESSENTIAL BUS
PYRO BATTERIES
BUS
BAT
(ESS. EMERG. O'RIDE)
AUX BATTERY
FEEDER CONTROL ASSY
CONTACTOR
PLSS BATTERY CHARGER
BCPK
OFF
FEEDER CONTROL ASSY
CONTACTOR
CB & LOGIC CONTROL NETWORK
(OFF)
ESSENTIAL BUS
FCA NO. 2 SIGNAL
(NON ESS SW POS)
FEEDER CONTROL ASSY
CONTACTOR
CB & LOGIC CONTROL NETWORK
DC
MONITOR BUS
GATE HI VOLT SELECT
DC CONTROL BUS
(ON)
FCA NO. 3 SIGNAL
(ON)
(AUTO POSITION)
SELECT CMDS
CONTROL PANEL SUBASSEMBLY
BUS
BAT
PYRO PWR SEL. SW
AUTO
MANUAL LOCKOUT
AUX BATTERY CONTROL
BUS CONTRO
SW

Figu

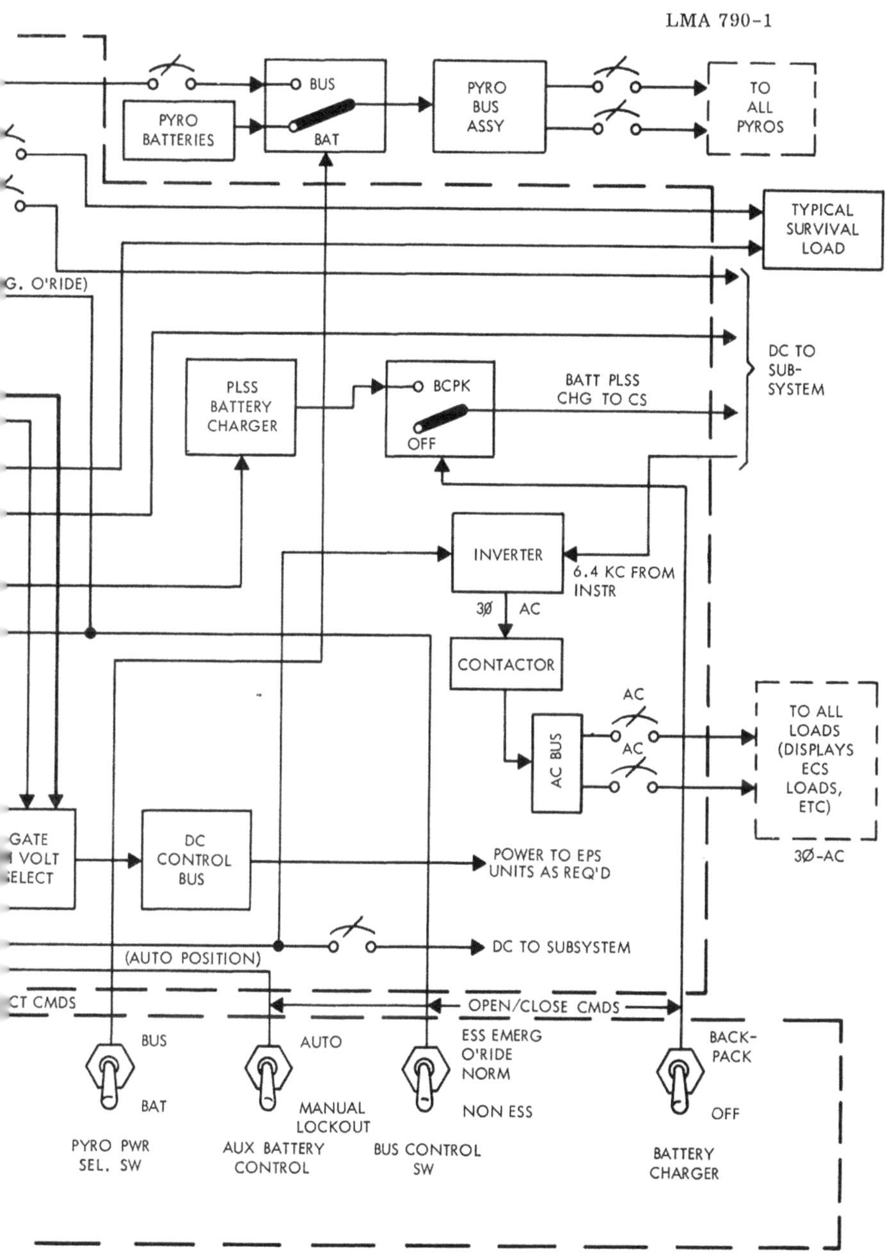

Figure 3-19. Electrical Power Subsystem Block Diagram

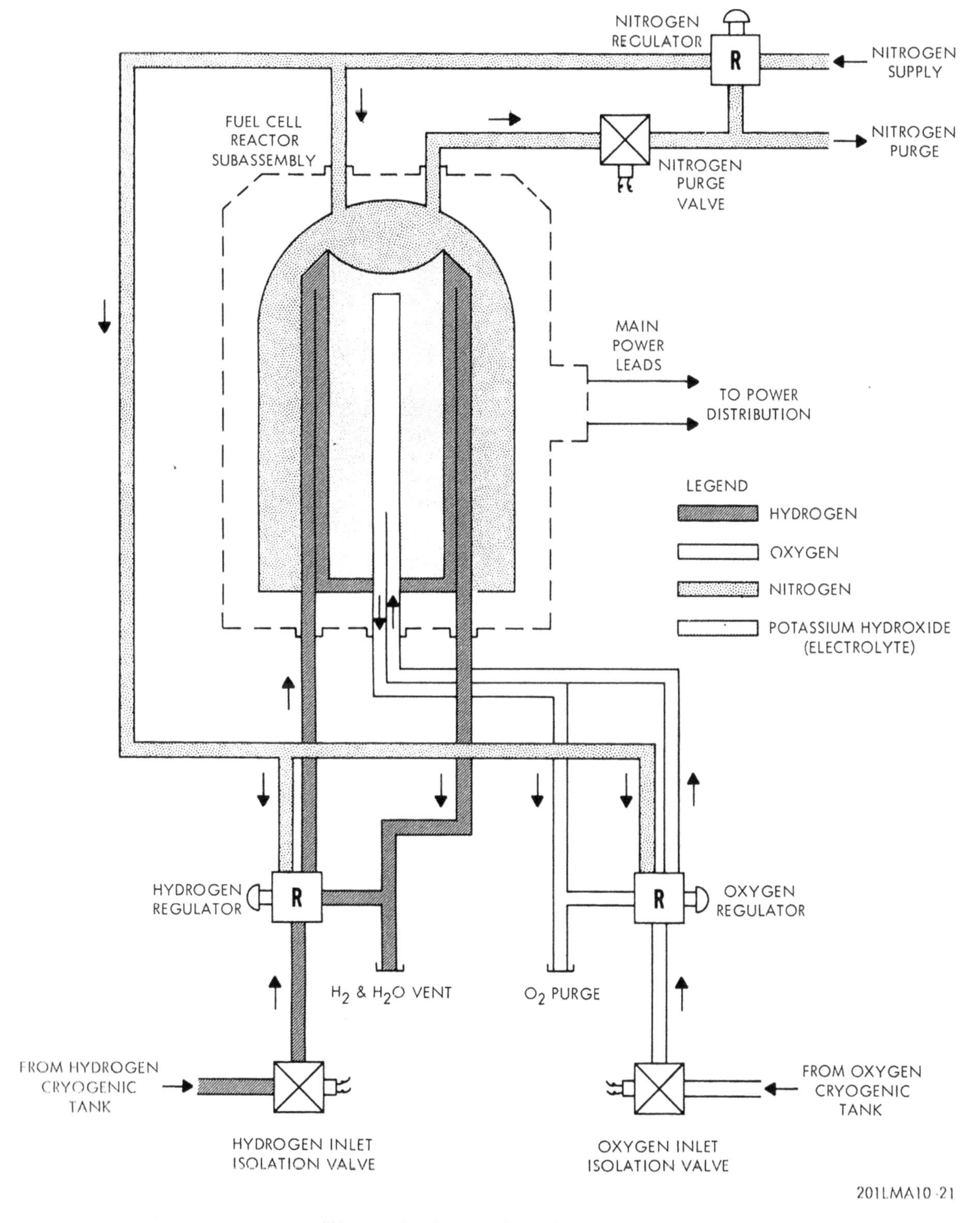

Figure 3-20. Fuel Cell Schematic

and the other pair sends a signal to the alarm system. The overcurrent relay is in series with the power output lead of the fuel cell assembly. When the current exceeds 150 amperes, the relay closes two pairs of normally open contacts. One pair sends a signal to the control relay and the other pair sends a signal to the alarm system. The overtemperature switch is operated by the temperature sensor in the hydrogen flow control valve. When an overtemperature condition exists, the switch closes and sends a signal to the control relay and to the alarm system. The gas pressure sensors are hermetically sealed potentiometric capsule instruments that monitor the pressure downstream of the oxygen, hydrogen, and nitrogen gas regulators. These instruments provide signals that can be used for telemetry and cockpit readout. The cell temperature sensors are platinum - resistance wire type instruments that sense the temperature of the fuel cell stacks. One sensor is mounted in each stack. The average of the two temperature signals is transmitted to the interface panel for telemetry and cockpit readout. The interlock system gives priority to the use of ground support equipment power (in place of fuel cell assembly power) when it is available.

3-84. AUXILIARY BATTERY.

The auxiliary battery used in the LEM is a silver oxide-zinc storage battery of sealed, wet construction. Each battery case is sealed and constructed to minimize leakage or loss of electrolyte for all conditions of battery operation in the earth, space, and lunar environments. The volume between the cover and the cells is filled with dry air and helium (the helium is a tracer gas) at a pressure of approximately 14.7 psi. The connector has redundant pins for the positive and negative terminals. The cell terminals are silver-plated brass.

3-85. ENVIRONMENTAL CONTROL SUBSYSTEM.

The Environmental Control Subsystem (ECS) consists of five integrated sections: atmosphere revitalization, oxygen supply and cabin pressure control, heat transport, water management, and cold plate. The major portion of the ECS is in the pressurized equipment compartment in the ascent stage. A glycol loop and a gaseous oxygen accumulator are in the ascent stage equipment bay section. Two water tanks are in the tankage section of the ascent stage, while a third (larger) water tank is in the descent stage.

The ECS (see figure 3-21) controls the oxygen in, pressurization of, ventilation of, and temperature of the cabin and the space suits worn by the two astronauts. It provides breathing oxygen for the astronaut's space suits and cabin; it limits the level of carbon dioxide and removes odors, moisture, and particulate matter from the oxygen breathed by the crew; and it automatically controls the temperature of the electronic equipment. The ECS also stores water for drinking, food preparation, and the Portable Life Support System (PLSS).

3-86. ATMOSPHERE REVITALIZATION SECTION.

The atmosphere revitalization section conditions and provides oxygen to cool and ventilate the space suits worn by the crew and monitors cabin oxygen recirculation and temperature. More specifically, the atmosphere revitalization section monitors the carbon dioxide level of the atmosphere breathed by the crew, removes odors and noxious gases from the crew atmosphere, removes foreign objects originating within the space craft, removes excess moisture from the cabin atmosphere, and provides control of the space suit gas temperature and the gas flow through the space suits.

The atmosphere revitalization section consists of two (one redundant) suit circuit fans, a suit circuit heat exchanger and water evaporator, and two (one redundant) water separators. In addition, the section contains a regenerative heat exchanger, a

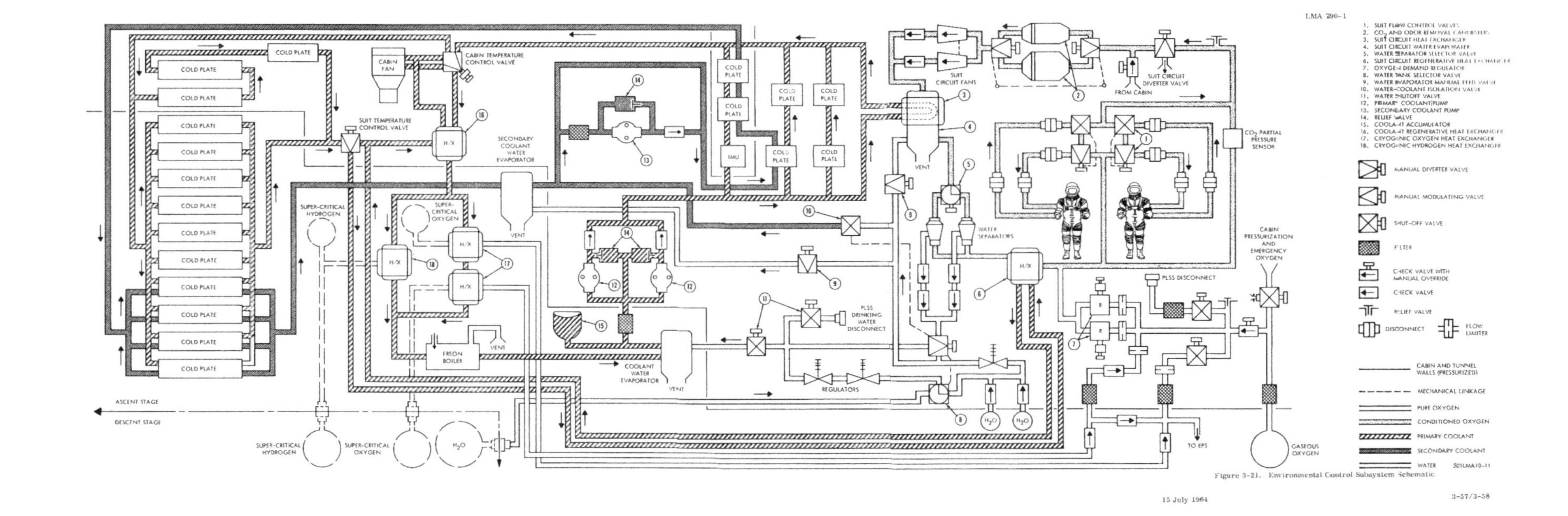

Figure 3-21. Environmental Control Subsystem Schematic

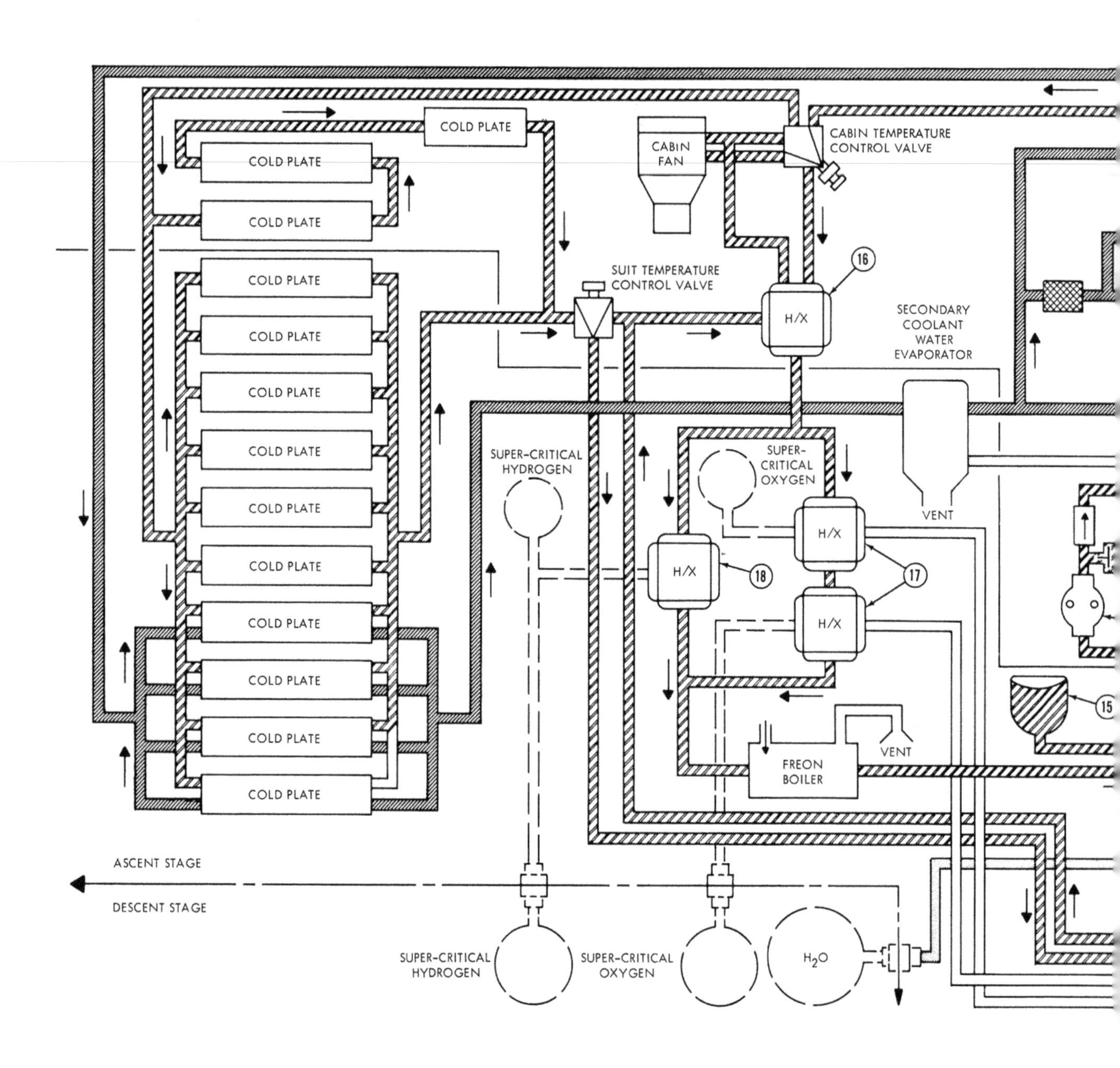
COLD PLATE
COLD PLATE
COLD PLATE
COLD PLATE
COLD PLATE
COLD PLATE
COLD PLATE
COLD PLATE
COLD PLATE
COLD PLATE
COLD PLATE
COLD PLATE
COLD PLATE
CABIN FAN
CABIN TEMPERATURE CONTROL VALVE
SUIT TEMPERATURE CONTROL VALVE
16
H/X
SECONDARY COOLANT WATER EVAPORATOR
SUPER-CRITICAL HYDROGEN
SUPER-CRITICAL OXYGEN
VENT
H/X
18
H/X
17
H/X
15
FREON BOILER
VENT
ASCENT STAGE
DESCENT STAGE
SUPER-CRITICAL HYDROGEN
SUPER-CRITICAL OXYGEN
H_2O

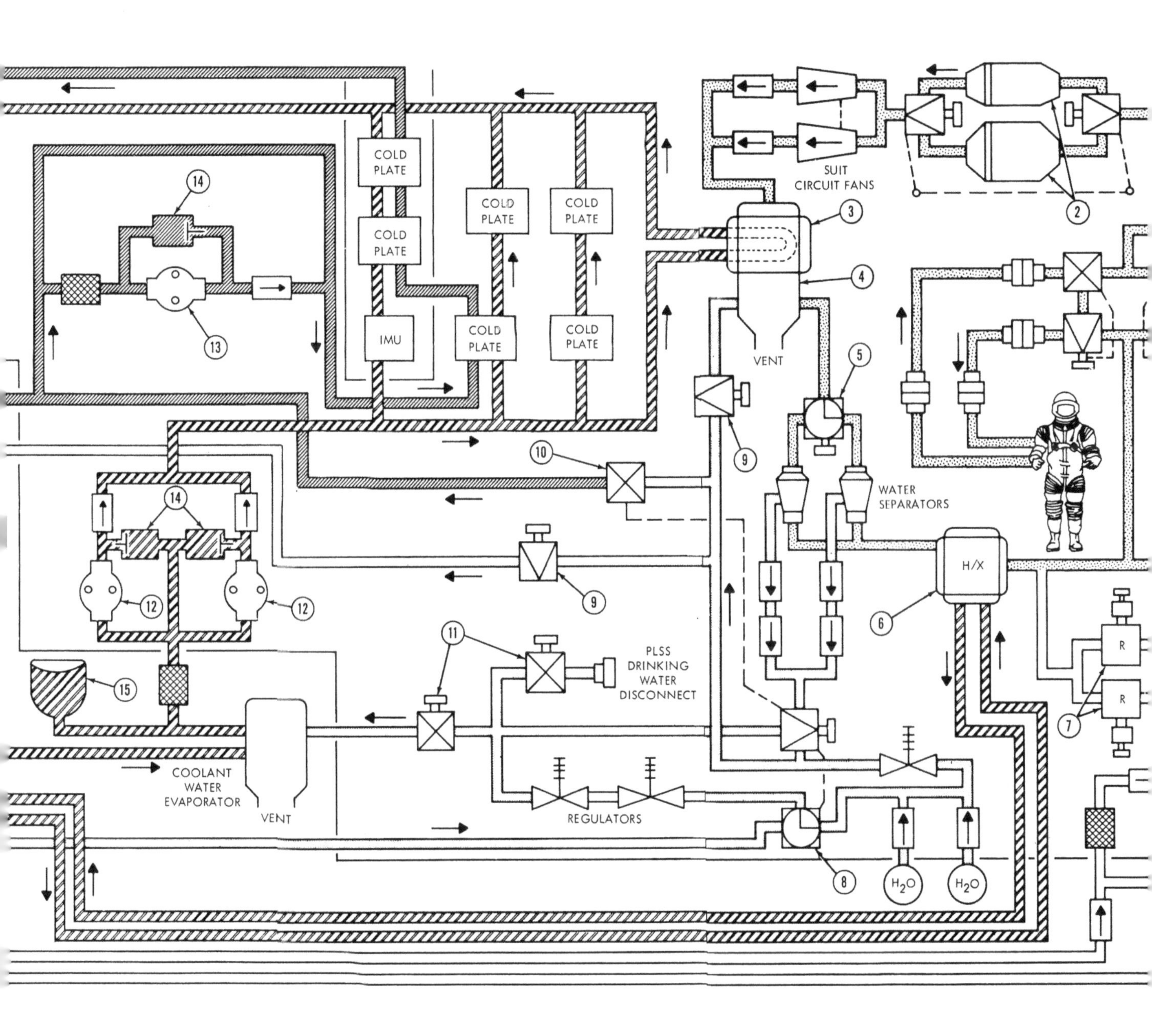
COLD PLATE
COLD PLATE
COLD PLATE
COLD PLATE
COLD PLATE
COLD PLATE
IMU
SUIT CIRCUIT FANS
VENT
WATER SEPARATORS
H/X
PLSS DRINKING WATER DISCONNECT
COOLANT WATER EVAPORATOR
VENT
REGULATORS
R
R
H_2O
H_2O
2
3
4
5
6
7
8
9
9
10
11
12
12
13
14
14
15

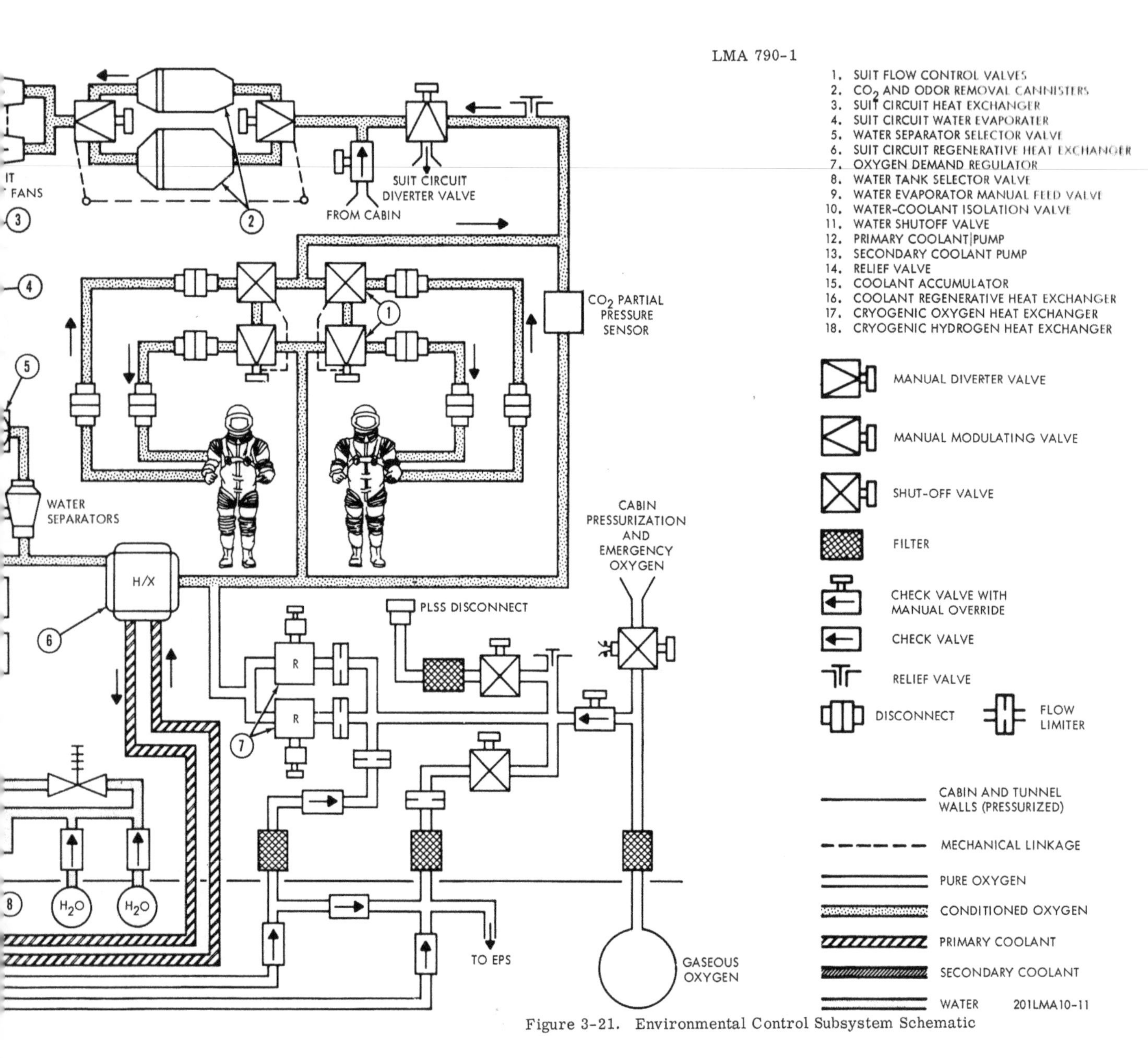

Figure 3-21. Environmental Control Subsystem Schematic

carbon dioxide partial pressure sensor, two carbon dioxide odor removal canisters, and relief valve, check valves and interconnecting tubing.

Oxygen from the oxygen supply and cabin pressure control section is circulated through the atmosphere revitalization section by one of two (one redundant) suit circuit fans. As either of the fans can maintain the required suit-circuit oxygen flow, only one fan is operated at a time. After leaving the fan, the oxygen passes through the suit circuit heat exchanger, which transfers excess heat from the oxygen to the heat transport section coolant. The suit circuit water evaporator removes the excess heat in the event of failure of the suit circuit heat exchanger. Any excess moisture that condenses when the oxygen passes through the suit circuit heat exchanger or suit circuit water evaporator is removed from the oxygen by one of two (one redundant) water separators. Each separator can meet the water removal requirements but depending on the position of the manually operated water separator selector valve, only one of the separators will function at a time.

Downstream from the two water separators is the suit circuit regenerative heat ex-exchanger, which allows the temperature of the oxygen to be manually controlled by the crewmember before the oxygen enters the space suits. Warm coolant from the heat transport section flows through the heat exchanger, transferring heat to the oxygen. The temperature of the oxygen is controlled by varying the flow of coolant through the heat exchanger; the astronaut manually controls this flow by the suit temperature control valve. In addition, the astronauts control their comfort by their individual flow control valve. The carbon dioxide partial pressure sensor maintains carbon dioxide at a safe partial pressure level. The oxygen then passes through one of two (one redundant) carbon dioxide and odor removal canisters, each consisting of a canister and replaceable cartridge, and once again flows into one of the two suit circuit fans where the cycle is repeated.

During open-faceplate operation (normal pressurization level), the suit-circuit diverter valve is opened to pass the entire oxygen flow from the atmosphere revitalization section into the cabin. This ensures that a sufficient amount of cabin oxygen is circulated through the atmosphere revitalization section to maintain the desired carbon dioxide and humidity levels in the cabin. In the event of a decompression of the cabin atmosphere, the cabin pressure switch provides a signal that automatically closes the diverter valve. The suit-circuit relief valve prevents overpressurization of the space suits. When space suit pressure is 4.4 psia or greater, the relief valve is fully open; when space suit pressure is less than 4.1 psi, the relief valve is fully closed.

Recirculation and temperature control of the cabin oxygen is provided by the cabin recirculation assembly. The assembly contains a fan that recirculates the oxygen and a cabin heat exchanger that automatically heats or cools the oxygen. Heat is transferred between the cabin oxygen and heat transport section coolant that flows through the cabin heat exchanger. The temperature of the coolant is controlled in the heat transport section (refer to paragraph 3-88).

3-87. OXYGEN SUPPLY AND CABIN PRESSURE CONTROL SECTION.

The oxygen supply and cabin pressure control section provides oxygen required by the atmosphere revitalization section and also supplies oxygen to refill the PLSS, used during exploration of the lunar surface. In addition, the section maintains cabin pressure by supplying oxygen at a rate equal to cabin leakage plus crew consumption, allows for cabin depressurization and subsequent pressurization by the astronauts, and maintains space suit pressure during unpressurized cabin operation.

The oxygen supply and cabin pressure control section consists of a gaseous oxygen accumulator, two (one redundant) oxygen demand regulators, various check valves, shutoff valves, and interconnecting tubing.

At the normal pressurization level, the pressure in the cabin and the space suits is maintained at 5±0.2 psia, which permits the astronauts to open their faceplates and remove their gloves. With the cabin depressurized, the space suits must be sealed and the pressure in the suits is reduced to the emergency level of $3.7^{+.02}_{-.00}$ psia.

The oxygen stored in the supercritical storage assembly is sufficient for six cabin repressurizations and six refills of the PLSS primary oxygen storage tanks, in addition to normal crew consumption and vehicle and space suit leakage. The gaseous oxygen accumulator is used in conjunction with normal oxygen flow for cabin repressurizations that require high oxygen flow rates. The self-sealing PLSS oxygen disconnect permits refilling the PLSS primary oxygen storage tanks.

Pure oxygen from the supercritical storage assembly in the descent stage passes through one of the cryogenic oxygen heat exchangers where it is warmed by the heat transport section coolant to make it compatible with the operation of components in the oxygen supply and cabin pressure control section and the atmosphere revitalization section. Two oxygen demand regulators (one redundant), each with a manual override control the delivery of oxygen to the atmosphere revitalization section in response to signals from pressure sensors. The cabin repressurization and emergency oxygen valve monitors the delivery of oxygen to the cabin for cabin repressurization or to slow the loss of cabin pressure in the event of a puncture in the cabin pressure shell, in response to signals from the cabin pressure switch. In addition, the valve has a manual override.

Overpressurization of the oxygen supply and cabin pressure control section is prevented by the oxygen pressure relief valve, which automatically relieves excess pressure by venting oxygen into the cabin. Overpressurization of the cabin is prevented by the cabin pressure relief and dump valve. This valve automatically relieves excess cabin pressure by venting oxygen overboard. The valve can also be manually operated, from inside or outside the cabin, to dump the cabin pressure overboard.

3-88. HEAT TRANSPORT SECTION.

The heat transport section of the LEM consists of two closed-loop systems - the primary and the secondary. Each of these systems circulates an ethylene glycol-water coolant to provide temperature control of the electronic equipment. In addition, the primary system provides temperature control of the oxygen circulated through the cabin and the space suits and to warm cryogenically stored fluids.

The primary coolant loop consists of two coolant pumps (one redundant), the cabin temperature control valve, coolant regenerative heat exchanger, Freon boiler, coolant water evaporator, coolant accumulator, coolant filter, various valves, and interconnecting tubing. The secondary coolant loop, used for cooling of critical equipment in the event of primary system failure and consists of a coolant pump, coolant water evaporator, filter, valves, and interconnecting tubing.

In the primary coolant loop, the coolant is circulated by one of the two coolant pumps. Each pump can provide normal flow and only one pump is operated at a time. After leaving the pump, the coolant flow divides, some going through the suit-circuit heat exchanger to transfer the heat from the oxygen to the coolant and some going through part of the cold plate section where it absorbs heat from the electronic equipment. The flow then divides between the regenerative heat exchanger and its bypass obtaining the required heat for the cabin heat exchanger from the regenerative heat exchanger.

The flow split is made by the cabin temperature control valve which is controlled by the cabin heat exchanger discharge temperature. The cabin heat exchanger discharge coolant temperature is maintained within a narrow range of temperature by the valve which in turn maintains the heat exchanger discharge temperature and cabin temperature within the required temperature range.

The coolant then passes through another parallel cold plate section where the coolant receives rejected waste heat from the electronic equipment. After passing the cold plate section, the flow is controlled to the suit circuit regenerative heat exchanger for suit heating by the suit temperature control valve. Warm coolant then flows in parallel paths through the cryogenic oxygen and hydrogen heat exchangers to warm the oxygen and hydrogen supplied from the super critical storage assemblies. Waste heat is removed from the coolant by the water evaporator. The water subliminator removes heat from the coolant by the sublimation process and its products discharged overboard. The coolant then flows through the coolant filter (which removes particles that could cause a malfunction) and into the coolant pumps continuing the cycle.

The coolant accumulator maintains pressure above the coolant vapor pressure in the heat transport section and accommodates volumetric changes of the coolant.

Ground support provisions are provided for by two sets of two self sealing quick disconnects for the primary and secondary circuits. Each set of fittings provide for supply and return of the GSE coolant and are used for coolant fill and draining and ground cooling during equipment and system checkouts.

The secondary circuit coolant pump which only operates during flight where the primary system has been lost, circulates coolant through the emergency (safe return) equipment - cold plates, water boiler and filter. The water boiler removes heat from the coolant by evaporation.

3-89. WATER MANAGEMENT SECTION.

The water management section of the ECS stores the water for the metabolic needs of the crew, vehicle cooling and, for two fills and four refills of the PLSS water tanks. The water management section consists of two water tanks in the ascent stage, one in the descent stage, water pressure regulators, check valves, shutoff valves, and interconnecting tubing.

The water tanks are pressurized prior to launch to maintain the required pumping pressure in the water tanks. The water tank in the descent stage supplies the water required up to the time of lunar launch. After lunar launch, water is obtained from the two smaller tanks in the ascent stage. In addition to water from the tanks, water from the atmosphere revitalization section water separators is used in the water management section. The self-sealing PLSS water disconnect permits filling and refilling the PLSS water tanks and delivering of metabolic water for drinking and food preparation.

3-90. COLD PLATE SECTION.

A cold plate is physically attached to each piece of electronic equipment requiring active thermal control. These cold plates, which make up the cold plate section, are single-pass heat exchangers. Coolant from the heat transport section passes through the cold plates and removes the waste heat from the electronic equipment.

3-91. CREW PROVISIONS.

Crew provisions sustain life conditions in an environment that is hostile to life. Without these provisions, a manned mission would be impossible.

3-92. SPACE SUIT.

The space suit is the basic item of the life-support system. It shields the crew-members from the thermal-vacuum environments of outer space. The astronaut is thereby able to leave the LEM in free space to perform various functions. The space suit also provides short-term emergency backup protection during the loss of cabin pressure. The materials that are used to make the space suit not only resist the abrasive and radiation environments of free-space and the lunar surface, but also resist suit damage during crew ingress-egress operations via the LEM passages.

3-93. SPACE SUIT ASSEMBLY.

The space suit is composed of various components and is called a space suit assembly (SSA). To accomplish a safe, manned mission the suit is constructed as an anthropomorphic, closed circuit, pressure vessel enveloping the entire crewman. Included in the SSA are biomedical and environmental sensors, telemetry and communications devices, and a portable life support system (PLSS), thermal coveralls, boots and gloves, the emergency oxygen subsystem and constant wear garments. With the PLSS attached to the suit, the astronaut has a livable envelope that can sustain his activities either within the LEM or in the lunar environment. A buddy system hose renders additional emergency capability between SSA's.

3-94. SPACE SUIT ASSEMBLY PERFORMANCE.

Inside the SSA's aluminized external layer, the crewmember is supplied a 100% oxygen atmosphere. The multi-layered SSA operates normally at 3.7 psi pressure and weighs approximately 60 pounds.

3-95. SPACE SUIT ASSEMBLY MOBILITY.

Space suit construction is dictated by mission objectives and the design features of the vehicle. The suit permits the astronaut to enter and leave unaided through either LEM passage during the mission. Therefore, manipulation of the feet, hands, legs, forearms, arms, head, and torso is possible. The astronauts will be able to walk, climb, crouch, and rise from a supine to a standing position. The use of tools, voice and telemetry communications, access to visual sightings (inside and outside LEM), and the reading of and subsequent use of proper spacecraft controls are included in the SSA's performance capabilities.

3-96. SPACE SUIT ASSEMBLY COMPONENTS.

3-97. Constant Wear Garment. The constant wear garment (CWG) is worn by each astronaut during all LEM operations. It is the SSA's inner-most component and serves to retain perspiration residue and body heat. The garment resembles long john underwear, and is made from cotton.

3-98. Pressure Garment Assembly. The pressure garment assembly (PGA) includes a helmet, a torso and limb suit, a pair of gloves, and lightweight boots which are an integral part of the torso suit. The pressure garment assembly items are worn during all LEM operations. Pressure leak-sealing capability is provided for contingencies. This capability extends to those crew-initiated situations that require puncturing the SSA. For example, should a crewmember's injury or illness require a hypodermic needle injection through the pressurized SSA the puncture in the pressure garment assembly is resealed after the hypodermic needle is removed.

3-99. Helmet. The primary function of the helmet is to protect the crewmember's head against high impact loads. With glare visors attached to the helmet's exterior, the crewmember can read vehicle and space data; the wide range of illumination environments necessitates the use of glare visors. Microphone and voice communications assemblies are contained within the helmet. Helmet construction is compatible with the crewmembers physiological needs and permits eating and drinking.

3-100. Torso and Limb Suit. The torso and limb suit is the anthropomorphic pressure vessel covering the crewmember's body and limbs, excluding the head, arms, and hands which are covered by the helmet, and gloves. Provisions for accepting the helmet, gloves, and boots completes the crewmember's enclosure within the normal 3.7 psi atmosphere. The torso and limb suit also has a re-sealing capability if punctured by a foreign object.

3-101. Gloves. The pressure garment assembly gloves are pressurized to extend the SSA closed pressure vessel to the crewmember's hands. The thermal gloves insulate the crewmember's fingers from thermal environments. When the pressure garment assembly is unpressurized, the astronaut can put on and take off the gloves with a minimum of effort and time. These gloves provide adequate finger dexterity in all pressurized environments. Each astronaut has another pair of gloves that are not a part of the pressure garment assembly. They are used for external activities.

3-102. Thermal Boots. Thermal boots provide thermal insulation, support, and protection. The sole and heel surfaces are designed to aid walking inside all modules of the LEM. Assembled to the limb and torso suits, the boots link the feet to the closed circuit pressurized envelope of the pressure garment assembly. As in the case of the gloves, the astronauts are furnished another pair of boots for use on the moon. Similarly, these boots are not part of the pressure garment assembly.

3-103. Thermal Garment. The thermal garments include a thermal suit, extravehicular gloves and extravehicular boots. These items, are not included in the pressurized atmosphere of the space suit but are worn over the pressure garment assembly. The main function of the thermal garment components is to provide insulation and protection against harmful radiation doses and thermal radiation.

3-104. Thermal Suit. The thermal suit provides thermal insulation from the lunar surface environment. Excluding the helmet and gloves, the reflective suit is worn over the portable life support system and the pressure garment assembly. It is a loose-fitting, two-piece, multi-layered, pajama suit covering with an aluminized outer coating. The metallic outer coating renders the suit reflective. The design of the suit allows the astronaut unaided, self-donning capabilities in the LEM.

3-105. Extra-Vehicular Gloves. Each astronaut is furnished with a pair of extra-vehicular gloves that are worn over the pressure garment assembly gloves. These gloves provide insulation protection and are worn during the stay on the moon. They may also be removed for short intervals for added dexterity. While wearing these gloves, the crewmembers finger dexterity to perform emergency and maintenance tasks, or to manipulate and erect mission task equipment is not hindered. Since the gloves are designed to be worn and taken off at any time, means are provided to attach the gloves to the SSA. A lanyard linking the gloves to the SSA sleeves is the arrangement presently utilized.

3-106. Extra-Vehicular Boots. A pair of extra-vehicular boots is provided each crewmember for wearing on the moon. These boots are worn over the lightweight boots, providing additional insulation. These boots, like the thermal suit have a reflective outer surface. The astronaut will be able to put on the boots unaided in the LEM prior to descending to the lunar surface.

3-107. PORTABLE LIFE SUPPORT SYSTEM

The portable life support system is a self-contained, (figure 3-22) rechargeable system providing limited-time life support for a crewmember exposed to conditions of extravehicular free space, a decompressed LEM, or the lunar surface environment. The backpack consists of various subsystems components providing assistance or operating modes for: Primary oxygen supply storage, contamination control, humidity control, pressure control, ventilation control, temperature control, recirculatory control, electrical power control, voice communications facilities, and telemetry transmission facilities. These functions are related only to the portable life support system backpack.

3-108. Primary Oxygen Supply. Crew oxygen provisions include two portable rechargeable pressure oxygen reservoirs supplying pure oxygen to satisfy body needs for normal and emergency situations. Compression of oxygen serves to reduce the size and weight of the reservoir which fits inside the portable life support system backpack. There is approximately four hours of oxygen available from each reservoir filling. The umbilical fittings used to charge the portable life support system reservoirs and to discharge its contents are common-usage CM/SM - LEM components.

3-109. Contamination Control. Contamination control includes the removal of explosive, noxious, nauseous, or toxic gases, solid particles and excessive moisture from the pressurized SSA's recirculatory system.

3-110. Humidity Control. A relative humidity of 40 to 70% within the temperature ranges of 66° to 70°F is maintained in the controlled normal and emergency environment of the pressurized SSA.

3-111. Pressure Control. Primary oxygen is used to maintain a steady operating 3.7 psi pressure to the SSA.

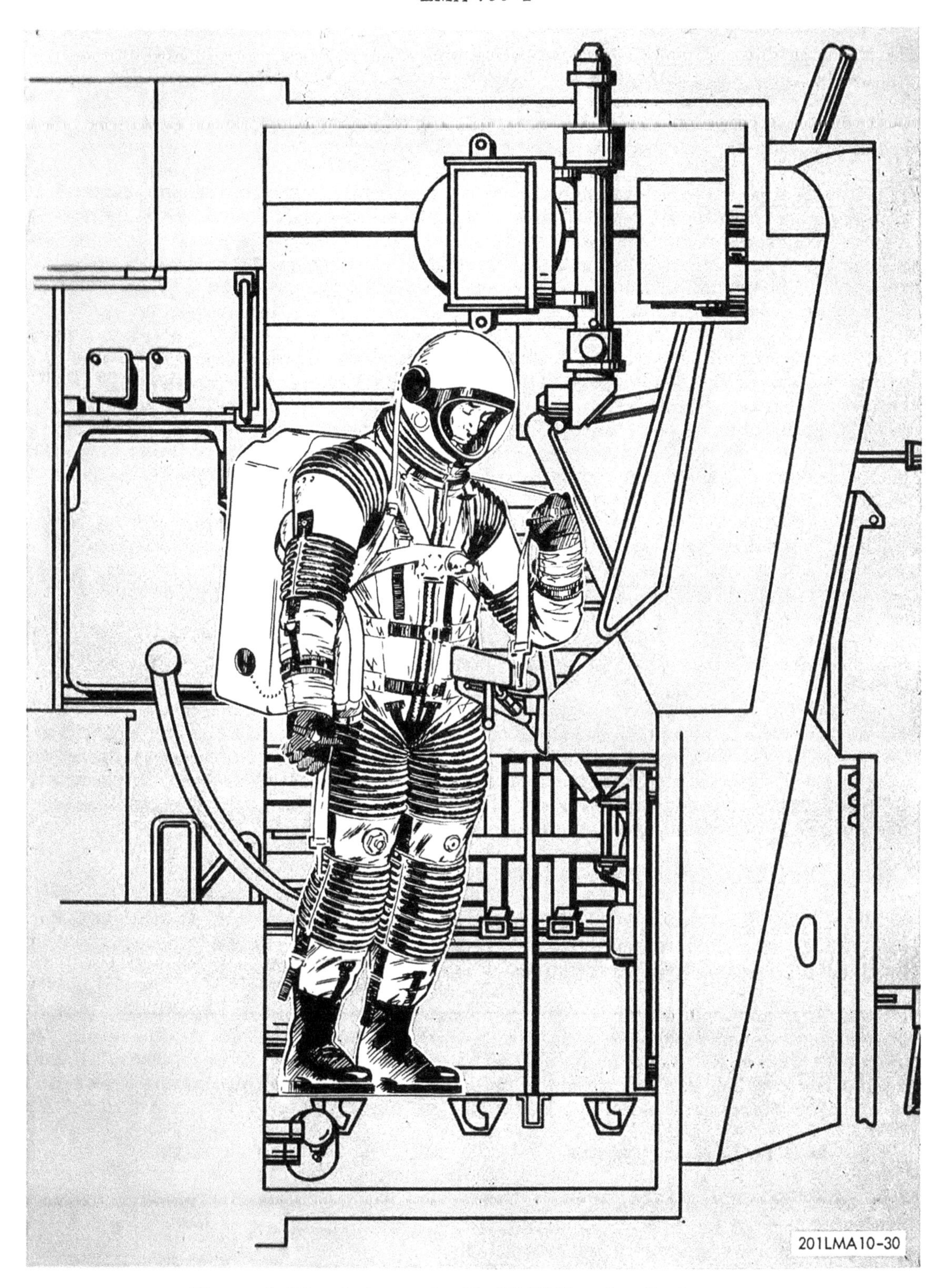

Figure 3-22. Astronaut with PLSS Donning Recharge

3-112. Ventilation Control. The ventilation or recirculatory system provides for the conditioning and recirculation of oxygen in the pressurized suit environment.

3-113. Temperature Control. A temperature range of 66° to 70°F is maintained in either a pressurized or unpressurized suit configuration.

3-114. Electrical Power. Electrical power is supplied by a rechargeable silver-zinc portable life support system battery capable of being recharged by the battery charger connected to the main electrical system in the LEM. An electrical umbilical is used to connect the portable life support system battery to the battery charger. The portable life support system is sufficient to power all SSA equipment for approximately 4 hours.

3-115. Voice Communications. A duplex subsystem is provided for simultaneous two-way communications between the crewmember on the lunar surface and the LEM, between two extravehicular crewmembers, (in SSA's), or between the crewmember in the SSA on the lunar surface and the Command/Service module transmitter and receiver. The communication system microphone and earphones are in the pressure garment assembly helmet. The antenna for the system is in the portable life support system backpack.

3-116. Telemetry Transmission. A telemetry subsystem transmits environmental and biomedical data to the Deep Space Instrumentation Facilities (DSIF) using the LEM subsystem as a relay link. Telemetry information is transmitted only on duplex as desired.

3-117. WAIST UNIT

The waist unit is the second component of the portable life support system and may be worn with or without the backpack. Provisions for electrical and electronic hardware and an umbilical connection to the torso and limb suit, using the same connectors available for the backpack are included. When the waist unit is worn, it is connected to the backpack. The waist unit contains portions of the emergency pressurization system, and emergency voice communications facilities.

3-118. BIOMEDICAL AND ENVIRONMENTAL SENSORS

3-119. Biomedical Sensors. The SSA's biomedical sensors gather physiological data for telemetry. A government furnished impedance pneumograph records respiration and cardiac impulses. Body temperature is also obtained for telemetering.

3-120. Environmental Sensors. The crewmembers monitor some of the SSA environmental sensors but other data are telemetered for ground monitoring. These measurements include suit inlet temperature and pressure, oxygen quantity, cumulative radiation dose, and elapsed time. An audible warning system informs the crewmember of low oxygen pressure and of excessive oxygen flow rate.

3-121. WASTE MANAGEMENT

In-suit waste management devices include provisions for removal of vomitus, urine and feces as well as flatus decontamination. These operations are possible under both pressurized and unpressurized conditions.

3-122. FOOD

Prepared foods suitable for eating in space environments is supplied to the crew-members. The menus include both liquid and solid foods. Food requirements criteria require food of adequate nutritional and caloric values and of low residue producing characteristics to minimize biological wastes. To eat while wearing the pressurized SSA, the astronaut uses the food valve adapter.

3-123. Food Packing. To prevent spoilage and to conserve space, dehydrated food is sealed in pliable plastic packages that resemble bottles. The end of the package is a circular tube that connects this package with food preparation and the food valve adapter. In order to eat the packaged food the cap of the package envelope is torn off, the water umbilical hose is attached to the package, and water is added to the dehydrated mix, resulting in a food mixture. The food package is then attached to a food valve adapter, which connects to the pressurized SSA helmet. The purpose of this adapter is toprovide a bridgefor the transfer of food from the container to the mouth. The crewmember then squeezes the pliable food container, to push the food from the container into the crewmember's mouth.

3-124. Food Valve Adapter. The food valve adapters make possible the transferring of food from the food packages to the inside helmet faceplate in front of the crewmember's mouth. Two adapters are provided, one for each astronaut.

3-125. Disinfectant Bottle. Since it is impossible to remove all food particles from the surfaces of the food valve adapters and food packages when the food is eaten, the food disinfectant bottle is used to prevent bacteria growth and decay in the residue particles. The bottle contains six ounces of disinfectant material.

3-126. OCCUPANCY AND RESTRAINT SYSTEM.

LEM crew stations positions are described as occupancy and restraint systems. The restraint system restrains the astronaut against impact loads while in motion in an optimum position for accomplishing his tasks. Unlike conventional seats, the LEM crew restraint systems confine the astronaut in an upright standing position during flight operations. The occupancy and restraint system includes a harness, arm restraints, waist-positioned pulley-reel assemblies, head restraints, and feet anchorage devices.

3-127. CREW STATIONS

Two occupancy and restraint positions are provided in the forward part of the main cabin. Each astronaut faces the main control and display panels (figure 3-23), and is in line with the triangular windows. Longitudinal geometry extends upward from the cabin flight deck to the upward cabin shell structure. Flexibility is designed in the crew stations to allow crewmembers to enter and leave their stations and to perform required flight duties.

3-128. Harness. Each astronaut will wear a parachute-type harness. Support is provided by pulley lines that connect the harness to the cabin structures. These lines have reels to enable manipulation in increasing or decreasing slack. Assemblies built in the harness support the torso.

3-129. Forearm Support. Two elongated U-shaped assemblies provide forearm support for the crewmembers. Each assembly has two concave curved slots in which the forearm is rested. The crewmembers, when leaning on these rests, have sufficient freedom of motion to allow operation of the vehicle controls.

3-130. CHECK LISTS

Check lists are provided for sequential operation procedures, crew tasks and crew action requirements.

3-131. FIRST AID KIT

A government furnished first aid kit makes possible minor treatment of crewmembers during the mission.

3-132. RADIATION DOSIMETERS

The radiation dosimeter indicates the exposure to radiation. Serious, perhaps critical, damage results if radiation doses exceed a predetermined level. For quick and easy reference, the astronaut has a radiation dosimeter mounted to his SSA. Measurements to the upper and lower body extremities may require additional devices be attached to other parts of the SSA.

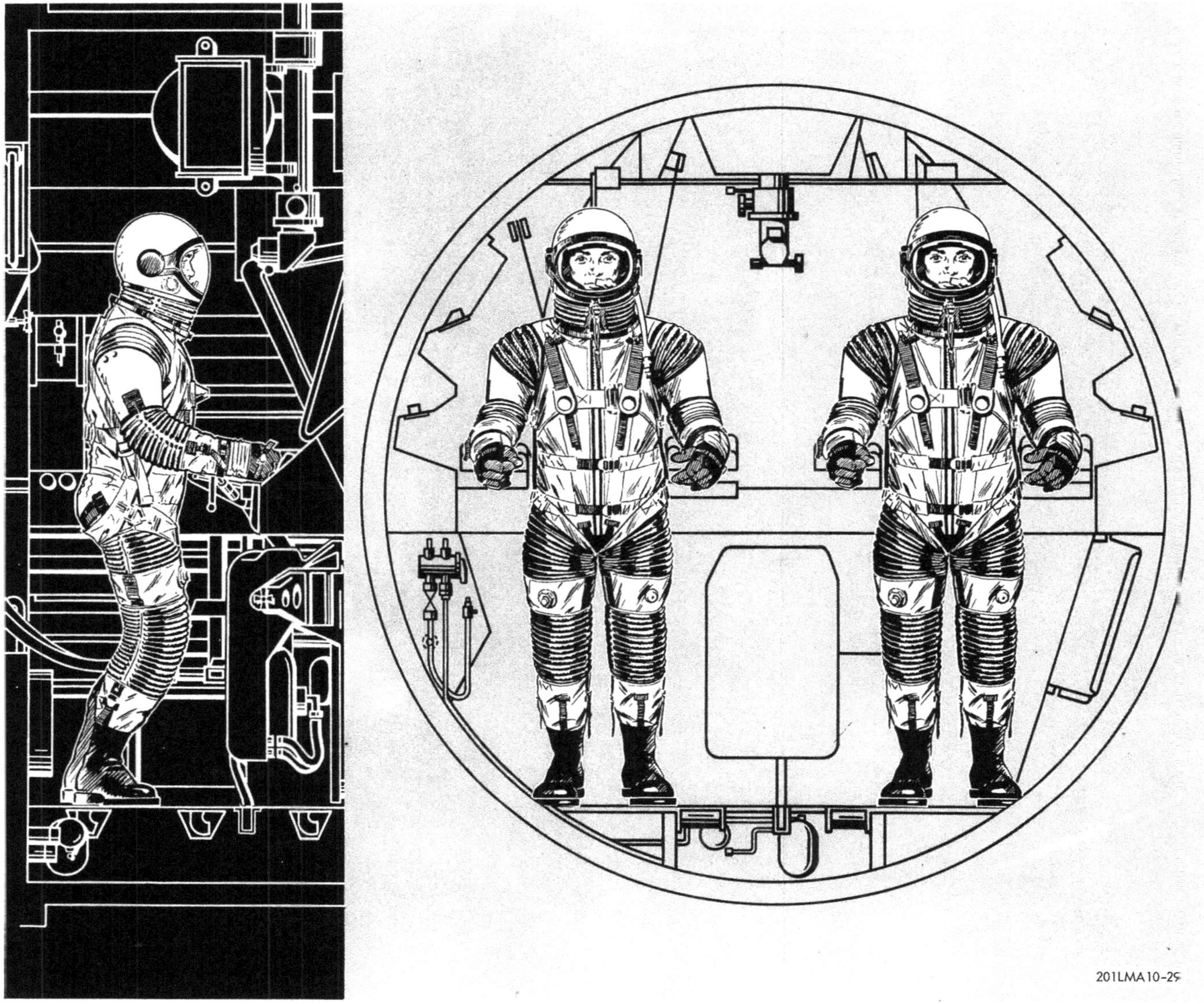

201LMA10-29

Figure 3-23. Astronaut Occupancy and Restraint

SECTION IV

PRELAUNCH OPERATIONS

4-1. GENERAL.

This section covers the prelaunch operation and facilities used for the LEM at the Atlantic Missile Range. The primary purpose of these procedures is to test and verify the functioning of each subsystem in all operational modes without removal of the equipment from the spacecraft.

4-2. PRELAUNCH TESTS AND OPERATIONS

Functional and verification tests are performed on the complete LEM system using fluid, mechanical, and electrical ground support equipment. These tests are automatically controlled by the Acceptance Checkout Equipment-Spacecraft (ACE-S/C).

Prelaunch verification tests are performed in chronological order at seven major facilities (figure 4-1): 1) Operations and Checkout Building, 2) Static Firing Site, 3) Weight and Balance Building, 4) Radar Boresight Range, 5) Vertical Assembly Building, 6) Launch Pad, and 7) Cryogenic Building.

4-3. ACCEPTANCE CHECKOUT EQUIPMENT - SPACECRAFT (ACE-S/C)

The Acceptance Checkout Equipment - Spacecraft (ACE-S/C) is a general-purpose spacecraft checkout system used for automatic, semiautomatic, or manually controlled acceptance testing and prelaunch testing of the LEM System. Its capabilities include the generation of test commands and stimuli, monitoring of spacecraft subsystem performance, conversion and processing of data, measurement of subsystem responses to test stimuli, diagnostic testing, and communication between the spacecraft and ACE-S/C controls. The ACE-S/C consists of the ACE-S/C Ground Station and the LEM Carry-On Equipment.

The general-purpose ACE-S/C Ground Station is adapted to the specific LEM subsystems and to the LEM servicing equipment by the LEM Carry-On Equipment, which consists of the two ACE-S/C Up-Links, the Communications Subsystem Checkout Equipment, the Radar Checkout Equipment, and an ACE-S/C Down-Link. One ACE-S/C Up-Link conditions and routes all ACE-S/C Ground Station test commands to their proper on-board equipment locations and supplies the necessary stimuli required by the subsystems undergoing test. The second ACE-S/C Up-Link performs much the same function for the ACE-S/C associated servicing equipments. The Communications Subsystem Checkout Equipment and the Radar Checkout Equipment perform special radiofrequency tests for the Communications and Navigation and Guidance Subsystems. The ACE-S/C Down-Link monitors the performance of the spacecraft subsystems and servicing equipments, and conditions and interleaves (time multiplexes) this data for transfer to the ACE-S/C Ground Station. The LEM Carry-On Equipment is described in greater detail in Section V.

Functionally, the ACE-S/C Ground Station comprises the following subsystems:

a. Pulse-Code Modulation (PCM) Subsystem, which prepares spacecraft checkout data for recording, displaying, and for transfer to the Computing Subsystem.

b. Computing Subsystem, which consists of a data processing computer, a digital command computer, and a common memory, all of which are housed in the ACE-S/C Computer Room. The data processing computer compares spacecraft checkout data from the PCM Subsystem with programmed tolerances and conditions it for CRT display. If an out-of-tolerance condition is detected, the appropriate indication is made to the test engineer at the malfunctioning subsystem's test console in the ACE-S/C Control Room. Diagnostic action can be initiated at the test engineer's discretion. All spacecraft test commands are generated in parallel form by the digital command computer and are converted to a serial format for transmission to the spacecraft.

c. Terminal Patch Facility, which uses coaxial hardlines and repeater-amplifiers to transmit data to and from the spacecraft and to remote displays.

d. Control Room, which contains the ACE-S/C displays and controls. Cathode ray tubes that display real-time test data from the Computing Subsystem in alphanumeric form are the primary display devices. These are supplemented by various analog and digital displays that receive their data directly from the PCM Subsystem. A summary of the complete spacecraft test is displayed to the system test engineer at his console in the ACE-S/C Control Room, and detailed breakdowns of the subsystem tests are displayed to the subsystem engineers at their Control Room consoles. The ACE-S/C controls enable three modes of testing. In the manual mode, subsystem engineers manually select individual, programmed, test commands. In the semiautomatic mode, programmed subroutines containing from one to several test commands are also selected manually. In the automatic-with-manual-override mode, test commands are generated under program control, in synchronization with a real-time clock.

The combined ACE-S/C facilities enable continuous processing of real-time spacecraft checkout data and complete test documentation for post-test analysis, with a minimum of interface cabling.

4-4. OPERATIONS AND CHECKOUT BUILDING

4-5. ASCENT STAGE.

4-6. Receiving Inspection. Upon arrival at AMR the LEM will be visually inspected for intransit damage. All assemblies, plumbing, connections, tie down hardware, electrical leads, and connections will be inspected for abrasions, breaks, cracks, loose fits or any detrimental condition which will require replacement or repair. Cryogenic and gaseous oxygen lines will be inspected for proper clearance from electrical lines and components. In addition, the data package will be reviewed to ascertain vehicle status.

4-7. Environmental Control Subsystem - External Leakage Check. The purpose of this test will be to verify that the external leakage rate of the heat transport section primary and redundant coolant loops are within specification tolerances. All valves will be set to the proper position and the leak detection unit connected. The blanket pressure will be vented to atmosphere and a leak tracer gas introduced in the coolant

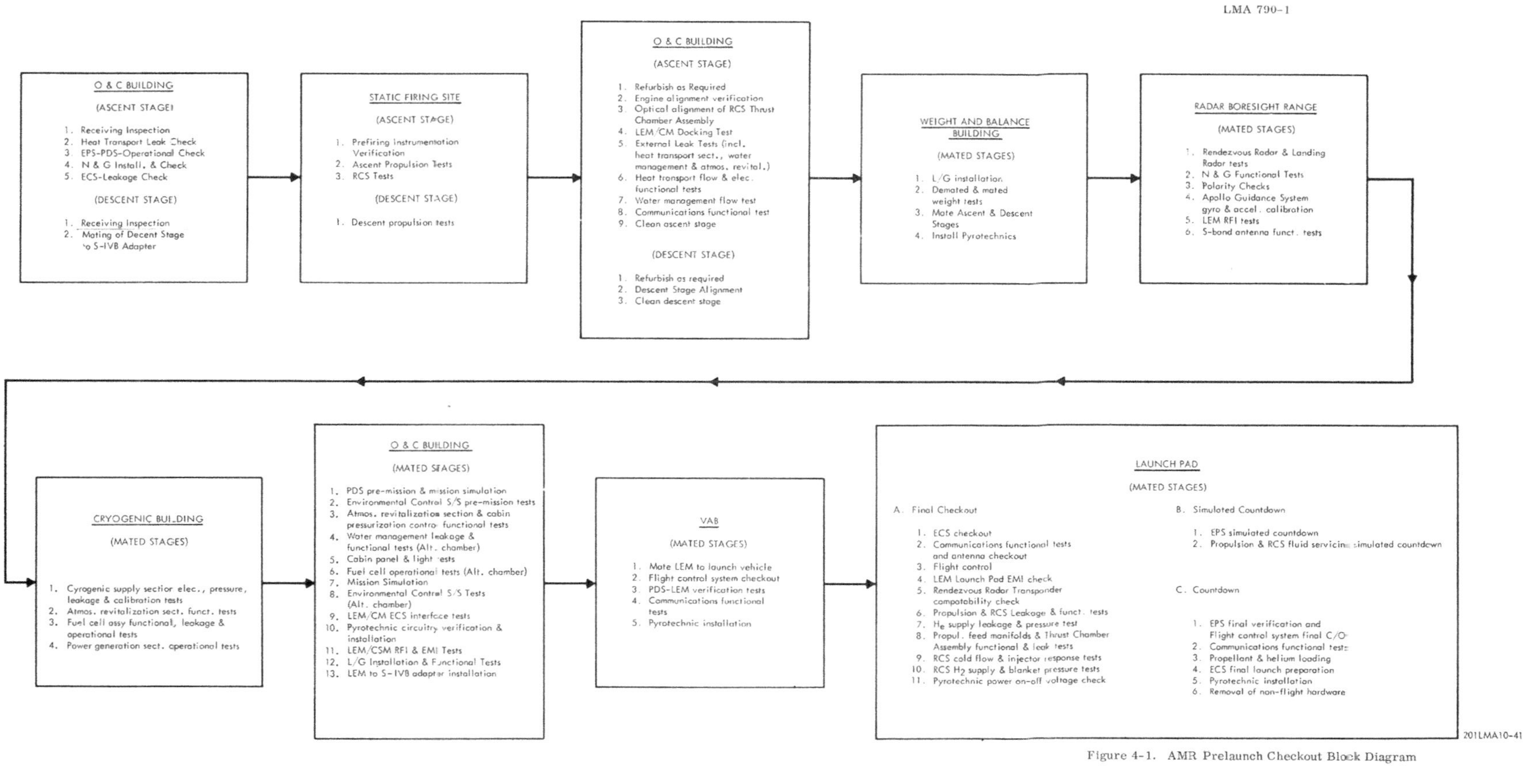

Figure 4-1. AMR Prelaunch Checkout Block Diagram

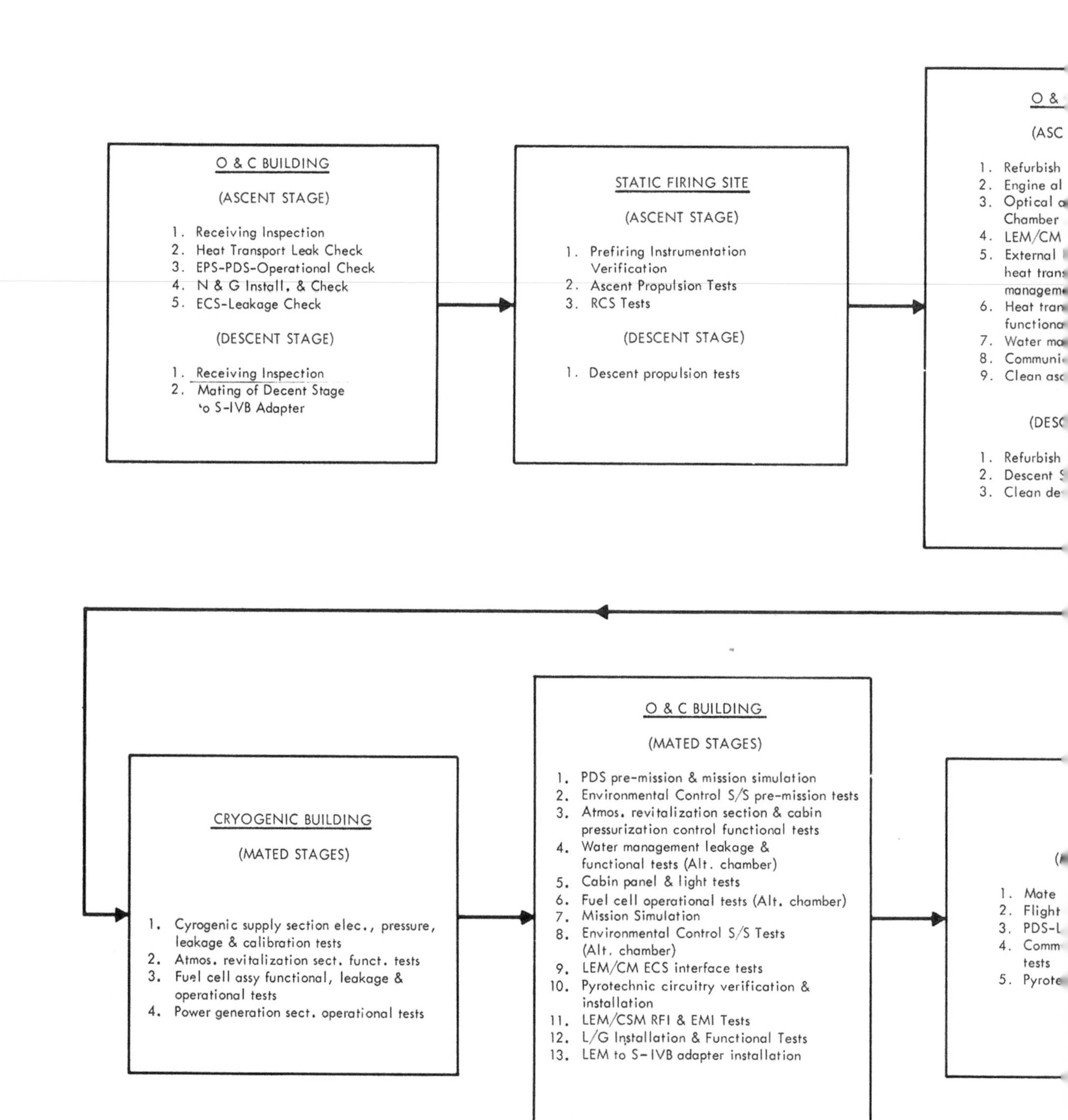

O & C BUILDING
(ASCENT STAGE)
1. Receiving Inspection
2. Heat Transport Leak Check
3. EPS-PDS-Operational Check
4. N & G Install. & Check
5. ECS-Leakage Check
(DESCENT STAGE)
1. Receiving Inspection
2. Mating of Decent Stage
to S-IVB Adapter
STATIC FIRING SITE
(ASCENT STAGE)
1. Prefiring Instrumentation
Verification
2. Ascent Propulsion Tests
3. RCS Tests
(DESCENT STAGE)
1. Descent propulsion tests
CRYOGENIC BUILDING
(MATED STAGES)
1. Cyrogenic supply section elec., pressure,
leakage & calibration tests
2. Atmos. revitalization sect. funct. tests
3. Fuel cell assy functional, leakage &
operational tests
4. Power generation sect. operational tests
O & C BUILDING
(MATED STAGES)
1. PDS pre-mission & mission simulation
2. Environmental Control S/S pre-mission tests
3. Atmos. revitalization section & cabin
pressurization control functional tests
4. Water management leakage &
functional tests (Alt. chamber)
5. Cabin panel & light tests
6. Fuel cell operational tests (Alt. chamber)
7. Mission Simulation
8. Environmental Control S/S Tests
(Alt. chamber)
9. LEM/CM ECS interface tests
10. Pyrotechnic circuitry verification &
installation
11. LEM/CSM RFI & EMI Tests
12. L/G Installation & Functional Tests
13. LEM to S-IVB adapter installation

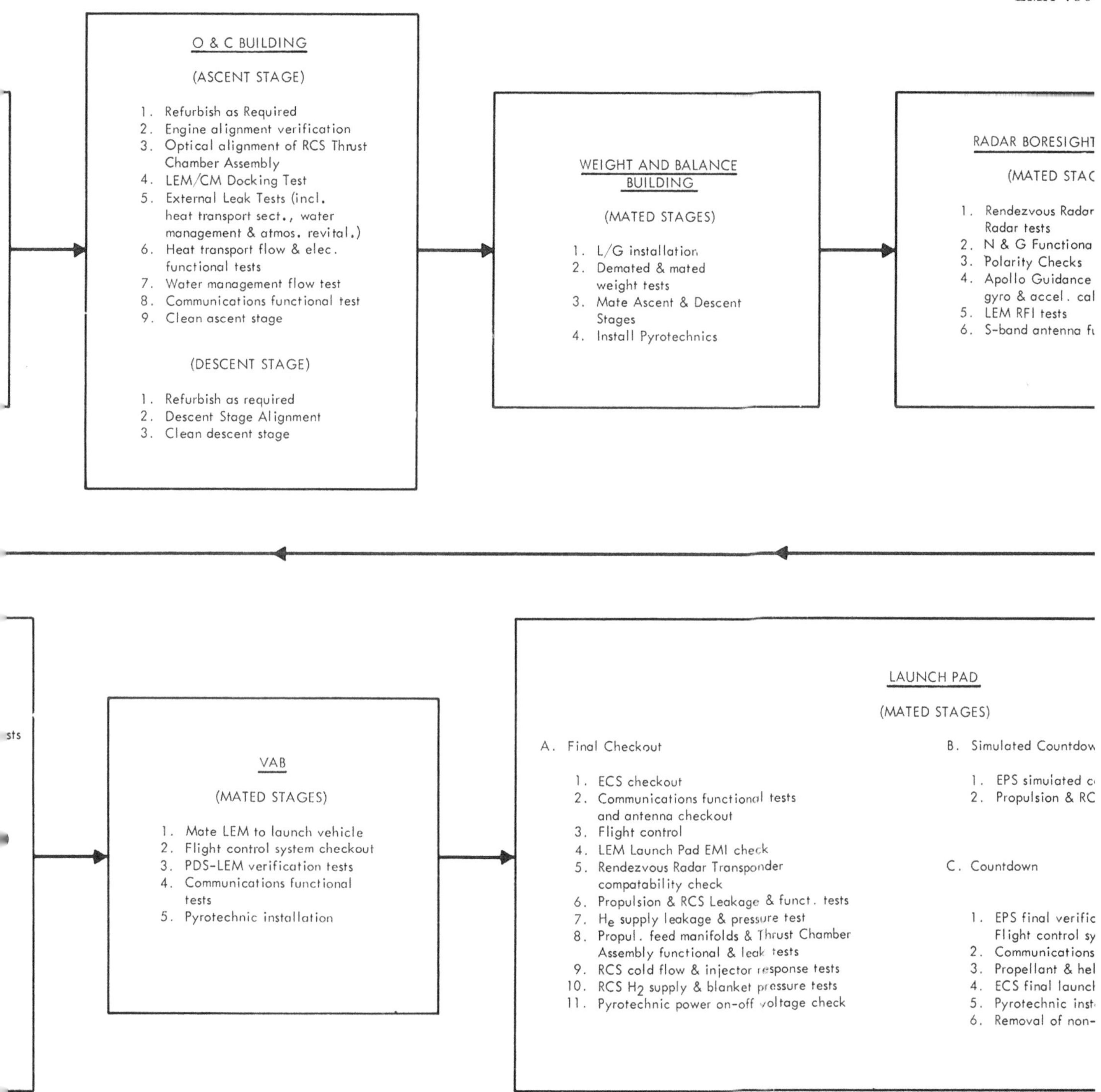

Figure 4-1. AMR Prelaunch Cl

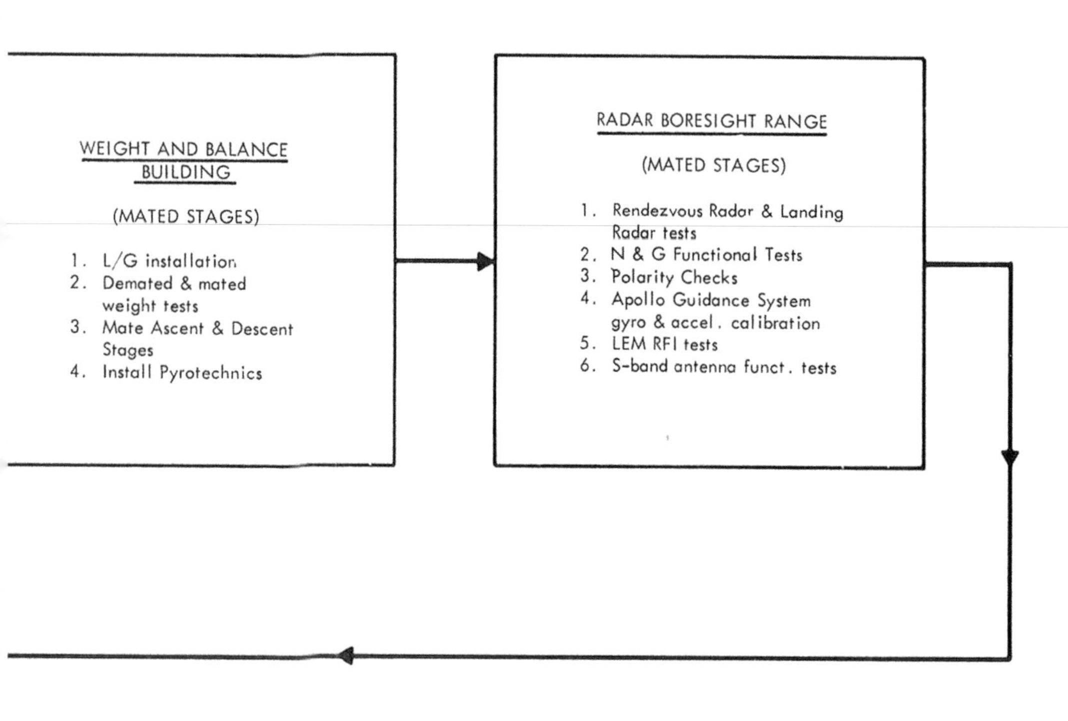

LAUNCH PAD

(MATED STAGES)

A. Final Checkout

1. ECS checkout
2. Communications functional tests and antenna checkout
3. Flight control
4. LEM Launch Pad EMI check
5. Rendezvous Radar Transponder compatability check
6. Propulsion & RCS Leakage & funct. tests
7. H_e supply leakage & pressure test
8. Propul. feed manifolds & Thrust Chamber Assembly functional & leak tests
9. RCS cold flow & injector response tests
10. RCS H_2 supply & blanket pressure tests
11. Pyrotechnic power on-off voltage check

B. Simulated Countdown

1. EPS simulated countdown
2. Propulsion & RCS fluid servicing simulated countdown

C. Countdown

1. EPS final verification and Flight control system final C/O
2. Communications functional tests
3. Propellant & helium loading
4. ECS final launch preparation
5. Pyrotechnic installation
6. Removal of non-flight hardware

201LMA10-41

Figure 4-1. AMR Prelaunch Checkout Block Diagram

loop. The section will be monitored for external leakage by utilizing a leak detector and monitoring for pressure loss. The redundant coolant loop will be subjected to the same test. At the completion of the test, the leak tracer vent will be to the atmosphere and the GSE water-glycol service unit will be connected to the coolant loop. The coolant loop will be purged evacuated and filled to mission level.

4-8. Electrical Power Subsystem - Distribution Section Operational Check. The purpose of this check is to verify adequate operation of the Power Distribution Subsystem, in both the manual and automatic mode while under a no-load condition with external power.

4-9. Navigation and Guidance Subsystem Installation and Post Installation Tests. The purpose of this test is to verify the Navigation and Guidance subsystem interface, and to ensure adequate operational continuity of the Navigation and Guidance Subsystem after reinstallation in the LEM.

4-10. Cabin and GSE Hatch Leakage Check. The purpose of this check will be to determine that the external cabin leakage rate is within specific tolerances. A GSE cabin test hatch will be installed on a docking tunnel. The other hatch will be closed but not latched. A GSE cabin leak test unit will be coupled to a cabin pressurization valve on the test hatch and the cabin pressurized to test pressure. A GSE pressure supply gage will be monitored to determine leakage rate. Compensation for temperatures will be made.

4-11. DESCENT STAGE.

4-12. Receiving Inspection. Upon arrival at AMR, the LEM will be visually inspected for intransit damage. All assemblies, plumbing, connections, tie down hardware, electrical leads, and connections will be inspected for abrasions, breaks, cracks, loose fits or any detrimental condition which will require replacement or repair. Cryogenic and gaseous oxygen lines will be inspected for proper clearance from electrical lines and components. In addition, the data package will be reviewed to ascertain the vehicle's status.

4-13. Mate Test-Descent Stage to S-IV-B Adapter. The objectives for this mating test are to ensure proper interface between the S-IV-B adapter and the descent stage, and to locate the adjustment plates of the adapter supports to level the LEM on the adapter for a positive fit. The lower SIVB adapter section will be in place on the adapter support stand and base in the LEM assembly and test area. The descent stage will be hoisted from the descent stage workstand and lowered into the lower section of the S-IV-B adapter. The adjustment plates on the Y-Y axis will be utilized to level the LEM to a common plane on the four support points. The LEM will be checked for fit to the adapter by attaching the straps on the Y-Y and Z-Z axis with nonexplosive bolts that will be used in place of the explosive units.

The holddown straps will be disconnected and the descent stage removed from the lower section of the S-IV-B adapter. The descent stage will be placed on the descent stage transporter and moved to the Static Firing Site.

4-14. STATIC FIRING SITE.

4-15. ASCENT STAGE.

4-16. Pre-firing Instrumentation Verification. This test will be conducted to ensure that instrumentation will function in support of the engine firing test requirements.

Additional checks of particular sensor calibrations will be integral in the engine checks. This test will be performed by activating the PCMTE and signal conditioners and verifying ambient readout of the temperature and pressure sensors. In addition, the pressure of inert gases stored in the fluid lines will be checked for leak rate determination. High- and low-rate switching of the PCM will be included in this test. The PCMTE verification will be provided by monitoring the calibration channels on the ACE.

Verification of accelerometer operation will be accomplished by stimulating the accelerometers with a portable vibrator. The On-Board Checkout Equipment (OBCE) and Caution and Warning Equipment (CWE) will be self-tested to assure operational capability as indicated on the appropriate cockpit displays.

Since the static firing will be unmanned, closed circuit TV will be used for display monitoring. The CCTV camera will be installed and checked out as part of this test.

4-17. Ascent Propulsion Test. The purpose of this test will be to verify the integrity of the propulsion subsystem. The check will include various leakage checks throughout the system, propellant loading, static-firing of the system, a post firing inspection sequence, and finally a cleaning and flushing operation of all oxidizer and fuel lines.

4-18. Reaction Control Subsystem Checks. As in the ascent propulsion test, the purpose of this test will be to verify the integrity of the Reaction Control System. The check will include thorough system leakage checks, propellent loading and pressure checks, static firing of the system, post-firing inspection, purging of system tanks, and cleaning of propellant total system.

4-19. DESCENT STAGE.

4-20. Prefiring Instrumentation Verification. This test will be conducted to insure that the instrumentation will function in support of the engine firing test requirements. Additional checks of particular sensor calibrations will be integral in the engine checks.

4-21. Descent Propulsion Checkout. The purpose of this test will be to demonstrate that the descent propulsion subsystem is functionally capable of performing a static firing, and to demonstrate the proper matching of the propellent feed system to the engine. The checkout will be identical to that of the ascent stage with minor exceptions; i.e., specific valve checks will be made during the actual descent propulsion checkout.

After completion of the static firing sequence, the ascent and descent stages will be returned to the Operations and Checkout Building for refurbishing as necessary.

4-22. OPERATIONS AND CHECKOUT BUILDING.

4-23. ASCENT STAGE.

During refurbishing various alignment procedures will be accomplished on the ascent stage. These will include engine alignment, optical, and instrumentation system alignment.

4-24. Docking and Leak Test - LEM To Command Module. The objectives of this test are to verify the fit between the LEM and the Command module at both the forward and overhead docking hatches, and to determine the leakage rate of the LEM when docked in either configuration.

The LEM and the Command module will be mounted on fixtures that allow the rotation of the LEM about the Y axis and relative movement of the two vehicles in the horizontal and vertical planes. The overhead docking surface of the LEM will be brought into contact with the docking surface of the Command module. The hooks of the locking mechanism of the Command module will engage the ring at the top of the LEM overhead docking tunnel. The LEM will then be pressurized above ambient. The docked hatch will be opened and the undocked hatch will remain closed. During the test, the hatch in the Command module will remain closed and LEM cabin pressure will be maintained in the Command module.

Pressure loss in the LEM cabin will be measured to determine the combined leakage rate of the LEM cabin and the docking interface. At the completion of this test, the LEM and the Command module will be separated.

4-25. External Leakage Test. The purpose of this check is to verify that the external leakage rate of the Water Management Section is within specified tolerances. A GSE leak detection test unit will be connected and a leak tracer gas supplied to the section under test. The section will be monitored for external leakage by utilizing a leak detector and monitoring for pressure decay.

4-26. Atmospheric Revitalization Section. The purpose of this check is to verify that the Atmospheric Revitalization Section leakage rate is within specification tolerances. After the various valves within the ARS are adjusted, a GSE leak detector test unit will be coupled to the section and pressurized to test pressure. Valves that require adjustment to permit test pressure to enter bypass loops will be adjusted during pressure buildup time. An external leakage test will be performed utilizing a leak detector test unit.

4-27. Cabin Leakage Check. This check will verify that the external leakage rate of the LEM cabin is within specific tolerances. The control valves within the Environmental Control Subsystem will be adjusted to the necessary position to perform this test. The upper and forward hatches will be closed and latched. A GSE cabin leak test unit will be coupled to the exterior of the LEM and the cabin pressure increased until the cabin pressure relief and dump valve cracking pressure is reached. Reseat pressure will be monitored and the GSE pressure supply gage will be monitored for leak rate.

4-28. Heat Transport Section - Electrical Function Test. The purpose of this check is to determine that the electrical portions of the primary and redundant coolant loop recirculation assemblies are functioning properly. A GSE water-glycol servicing unit will be coupled to the primary coolant loop supply and return line quick disconnects. The section will be topped off to mission level. External electrical power will then be applied to the section; however, the section circuit breaker will not be energized. After the appropriate controls within the coolant section have been adjusted, the Heat Transport Section will be energized and indicating lights checked for verification of operation. The coolant Automatic Control switch will be cycled manually and automatically by creating pressure drops across the recirculation assembly with the GSE service unit. Internal displays, in addition to ACE monitored pressures, will be used to check that operation of the system is satisfactory.

4-29. Heat Transport Section - Flow Test. The purpose of this test will be to verify that the components of the Heat Transport Section are functioning within acceptable limits. The Heat Transport Section pumps will be energized and automatic pump control will be confirmed with both manual and automatic switching. Verification of flow through the primary coolant loop will be monitored by recording the differential pressure across the primary coolant loop recirculation assembly and redundant coolant loop will be monitored by recording the water glycol pressure indicated on the main instrument panel.

4-30. Water Management Section - Flow Test. The purpose of this check will be to verify satisfactory operation of the fluid flow portions of the water management section. Various valves within the Water Management Section that will be affected by this test will be adjusted to meet requirements. Flow to the Heat Transport Section primary coolant loop water evaporator will be verified by allowing water to flow to the evaporator. Flow to the Heat Transport Section redundant coolant loop water evaporator and to the suit circuit water evaporator will also be verified. Flow to the Portable Life Support System water quick disconnect will be monitored. The time required to fill a test bottle, coupled to the section's quick disconnect, to a given point will verify the flow requirements at this point. All disconnects will be monitored for leakage.

4-31. Communications Functional Test. This check performs functional checkout operations of the communications system. This will include a check of the normal communications equipments, the back-up and emergency communications equipment, and the TV/Telemetry/Voice/Biological Medical data transmission facilities.

4-32. DESCENT STAGE PROPULSION CHECKOUT.

The purpose of this check is to verify the engine thrust alignment and gimbal drive operations. The engine alignment to the vehicle thrust axis will be verified optically, and the operation of the engine gimbal drive will be verified.

4-33. ASCENT AND DESCENT STAGE CLEANING.

To assure that foreign matter is removed from the interior of the ascent and descent stages before the final mating, the stages will be mounted on separate cleaning positioners and rotated about the Y and Z axes as necessary to allow a thorough cleaning of the interior. Upon completion of this operation, the ascent and descent stages will be transferred to the Weight and Balance Building.

4-34. WEIGHT AND BALANCE BUILDING.

4-35. LANDING GEAR INSTALLATION.

At this time it is necessary to install the LEM landing gear on the descent stage prior to the weight and balance tests.

The descent stage handling dolly with the descent stage handling dolly adapter installed will be in place in the Weight and Balance Building. The landing gear will be in the Weight and Balance Building ready for installation on the descent stage. The descent stage will be hoisted from the descent stage transporter on to the descent stage handling dolly. The landing gear will then be installed on the descent stage in the extended position. The main and secondary struts on each of the four legs will be installed individually on the descent stage and then connected to each other. The foot pads will be installed on the primary struts.

4-36. DESCENT STAGE WEIGHT TEST.

The purpose of this test is to verify the dry weight and lateral location of the center of gravity for the demated descent stage with extended landing gear.

The weight and balance fixture and equipment will be prepared for receipt of the descent stage in the Weight and Balance Building. This will include calibration checks on the load cells. After inventory and removal of all non-flight hardware, the descent stage will be hoisted from the descent stage handling dolly and placed on the LEM weight and balance fixture. The dry weight and Y-Z center of gravity location will be obtained by lowering the upright LEM onto the weight and balance fixture. Which contain three load cells that will bear against three hard points on the descent stage. There will be three weighings of the descent stage. Following the weighing, the descent stage will be hoisted from the weight and balance fixture and placed on the descent stage handling dolly for the mate test with the ascent stage prior to complete LEM weighing.

4-37. ASCENT STAGE WEIGHT TEST.

The objectives of this test is to verify the dry weight and lateral location of the center of gravity for the completely assembled ascent stage.

The weight and balance fixture and equipment will be prepared for receipt of the LEM ascent stage. After an inventory and removal of all nonflight hardware, the ascent stage will be hoisted from the ascent stage transporter to the weight and balance fixture and weighed three times.

4-38. MATE TEST - ASCENT TO DESCENT STAGE

The purpose of this test is to mate the ascent stage to the descent stage prior to weighing the completely assembled LEM. The descent stage will be on the descent stage handling dolly in the Weight and Balance Building, prepared for mating with the ascent stage. The ascent stage will be hoisted from the weight and balance fixture, placed on the descent stage in the descent stage LEM handling dolly, and the four attachment points between the ascent and descent stages will be bolted and secured. This is the final mating of the stages prior to launch, and explosive bolts will be used for the mating. The interstage umbilicals and hardlines will also be connected.

4-39. PYROTECHNICS - MECHANICAL INSTALLATION.

The objective of this operation will be to mechanically install the required pyrotechnic devices. A bridge wire continuity and electrical isolation test will be performed on all EED's in the Endurance Test Laboratory utilizing portable pyrotechnic continuity and electrical isolation testers to ensure that no stray voltages are present. The pyrotechnic devices will be installed upon completion of these tests, but electrical connections will not be made at this time. Instead, shorting connectors will be on all EED's electrical connections, until electrical connections are made. Positive identification as to location of each device installed in the LEM will be performed at this time.

4-40. RADAR BORESIGHT RANGE.

4-41. RENDEZVOUS RADAR/TRANSPONDER AND LANDING RADAR CHECKS.

The operation of the Rendezvous Radar, the Transponder, and the Landing Radar will be verified by: (1) optical and electrical boresighting of the Rendezvous Radar and Landing Radar antennas, (2) antenna voltage-standing-wave-ratio checks, (3) alignment checks of the Alignment Optical Telescope, (4) transmitter and receiver functional checks, (5) functional checks of the LEM Transponder, and (6) compatibility checks between the Rendezvous Radar and the remainder of the Navigation and Guidance Subsystem, which are performed while the Rendezvous Radar tracks a simulated target, to ensure the accurate transfer of data. The GSE used in performing these checks will include ACE-S/C.

4-42. RADIO FREQUENCY INTERFERENCE CHECKS.

Using ACE-S/C and other GSE, these checks will monitor the rf-sensitive LEM equipment under the following conditions: (1) with all LEM rf-radiating devices in operation, and (2) with the Tracking Range radiating devices pointing at the LEM. The latter check will subject the LEM to the worst possible external radiation conditions. The ability of rf - sensitive LEM equipment to reject internally-caused and externally-caused radio frequency interference will be verified.

4-43. S-BAND STEERABLE ANTENNA FUNCTIONAL TEST.

This test will verify the ability of the S-band steerable antenna to track a moving infra-red radiation source. During this test, the antenna will be aligned for maximum received signal strength, as indicated by "S-meter" readings.

4-44. POLARITY CHECKS.

These checks ensure the correct polarity of the attitude references, control systems, Propulsion and Reaction Control Subsystems, and displays. They are performed in the Attitude Hold mode. Before each polarity check, the Inertial Measurement Unit and the Abort Guidance Section are aligned to the LEM body axes. The LEM is tilted in each direction of pitch, roll, and yaw, and the following parameters are measured: LEM Guidance Computer attitude error signals, Inertial Measurement Unit and Abort Guidance Section accelerometer outputs, rate gyro outputs, RCS jet actuation signals, descent engine gimbal control signals, and attitude and attitude error displays.

4-45. CRYOGENIC BUILDING.

4-46. CRYOGENIC STORAGE SECTION LEAKAGE TEST AND PRESSURE DEVICE CALIBRATION.

These tests will be made to verify that the Power Generation Section, and hydrogen, oxygen and nitrogen storage and distribution components do not leak in excess of allowable limits, and will also verify that pressure transducers and switches in each storage section are functioning according the specification.

The cryogenic supply sections will be pressurized with gaseous helium, and leak checks will be made using a helium leak detector. The leak test will be made at the maximum allowable pressure for the oxygen and hydrogen storage sections.

During the Cryogenic Supply System pressure buildup, the transducer output will be monitored and its readouts compared with the actual tank pressure throughout the complete range of operation. The tank pressure will then be varied to obtain the open and closed actuation points of the pressure switch. The transducer and switch operation tests will then be repeated to verify the consistency of the developed data.

4-47. FUNCTIONAL TEST - ATMOSPHERE REVITALIZATION SECTION.

A series of tests will be performed on the Atmosphere Revitalization Section to ensure that it functions as an integral portion of the Environmental Control Subsystem. First of all, a GSE electrical resistor will be wired in series between the power supply and cabin fans. A suit loop stimuli generator to provide metabolic heat, water vapor, and CO_2 will be installed and cabin pressure and suit circuit will be evacuated. A vacuum will also be pulled in the primary coolant loop water evaporator of the heat transport section in order to provide the desired flow from the tanks of the water management section. The cabin and suit circuit fans will then be energized and the entire system subjected to various modes of operation to check various system configurations. Valve positions and temperature readings will be monitored. The lithium hydroxide cartridges will then be removed and CO_2 will be introduced into the system at controlled rates by the suit-loop stimuli generator to check the high-level-CO_2 indicator flag operation. The cartridges will then be replaced in the primary and bypass loops, section leakage will be measured, and the CO_2 canister selector valve will be actuated to verify that flow can be diverted from the primary loop to the bypass loop without obstruction. The partial pressure of the CO_2 will be monitored during this test. The water separator selector valve will then be positioned midway between selection detents to obtain a simulated pressure drop in the separators, and the operation of the flag position indicator will be checked. The primary coolant loop water evaporator coolant outlet temperature and differential pressure will be monitored to verify proper operation. The temperatures and pressures in the gaseous oxygen accumulator and water management quantity, temperatures, and regulator discharge pressures will also be monitored.

4-48. FUNCTIONAL AND LEAK TESTS ON CRYOGENIC SUPPLY SECTION AT OPERATING TEMPERATURES.

Supercritical reactants will be introduced into the cryogenic supply section and maintained at maximum operating pressures to verify that the CSS does not leak in excess of design limits and to check the operation of associated components. The interconnecting lines and stage disconnects will be included in the leak checks.

Oxygen and hydrogen supplies will be tested independently; each tank will be filled and sealed. Pressures will then be raised by using tank heaters, and both tanks will be leak checked. Component leak checks will be made by observing flow rates and pressure losses, and performed at varying pressures, with the final check made at the system relief pressure. Relief valve tests will be initiated slightly below relief pressure, and pressure will then be gradually raised by intermittent operation of the tank heater until the relief valve opens. Relief and reseat pressures of the valve will be checked for conformance to specification, after which the external leak checks of the system will then be repeated.

4-49. POWER GENERATION SECTION FUNCTIONAL AND ELECTRICAL TESTS.

Functional and electrical tests of the fuel cell assembly components will be made to verify that the circuitry and components function according to specification and that the maximum allowable leak rate is not exceeded. Immediately following completion of the fuel cell tests, a power generation section test will be made while the fuel cell assembly is still in operation in order to verify that the PGS performs within design limits.

4-50. OPERATION AND CHECKOUT BUILDING.

4-51. PDS VERIFICATION - PREMISSION AND MISSION SIMULATION CHECKS.

These procedures will verify the auto/manual operation of the PDS with external power in the chamber (at ambient pressures), and to verify the auto/manual operation of the PDS with fuel cells in the decompressed chamber. The power distribution system will be operated in the internal fuel cell power mode with controls positioned to auto-operation/normal mode and bus voltages checked. With the altitude chamber pressurized, auto operation on fuel cells with power will be checked. The voltage and frequency of the inverter bus will be checked. Inverter operation will also be checked by monitoring the a-c buses. Mission simulation will be accomplished with the fuel cells activated. With PDS controls in the auto-operation/normal mode, circuit breakers will be used for distribution of power to the subsystems for mission simulation checks. All buses will be monitored for voltage and current readings.

4-52. ENVIRONMENTAL CONTROL SUBSYSTEM - ATMOSPHERIC REVITALIZATION SECTION CHECK.

This test will verify that the electrical components of the Atmospheric Revitalization System are performing within specified tolerances. A GSE electrical resistor will be connected in series between the power supply and the cabin fans to prevent the fans from overheating when operating at 14.7 psia. A means of reducing the cabin pressure switch will be provided. A suit loop stimuli generator will be installed in the space suit assembly position. One or both O_2 in flow valves will be set to normal position, the suit circuit diverter valve will be set to normal open position and the cabin fans energized. The cabin pressure switch will be reduced to pressure of approximately 4.7 psia. The section will be monitored to verify that the suit-circuit diverter valve closes, the cabin repressurization and emergency O_2 valve opens, the master caution light flashes and the caution and warning annunciator light assembly glows. The pressure on the cabin pressure switch will be further reduced to approximately 3 psia. The cabin fans will be monitored to verify that they turn off. At this same pressure, both O_2 inflow valves will be set to egress mode. Verification will be made to assure that the cabin pressurization and emergency O_2 valve closes and the master caution light and caution and warning annunciator light assemblies go out. While maintaining 3 psia at the cabin pressure switch one or both O_2 inflow valves will be set to normal mode of operation. The cabin repressurization and emergency O_2 valve will be monitored to verify that it opens and the master caution light and caution and warning light annunciator light assembly will light. At pressures above 3 psia, the cabin fans will be monitored to assure that they are operating. Pressure at the cabin pressure switch will be increased to approximately 5 psia, the cabin repressurization and annunciator light assembly will go out. At this pressure, and with the O_2 inflow valves in normal mode, the cabin fans will be operated and the suit circuit diverter valve will open. Both O_2 inflow valves will be set to egress mode. The closing of the suit-circuit diverter valves closes and the stopping of the fans will be verified.

4-53. SUIT CIRCUIT FAN CONTROL CHECK

A GSE electrical resistor will be connected in series between the power supply and motor to prevent the suit-circuit fan from overheating. The normal fan start procedure will be followed. Adjustment of a variable resistor will be made until automatic switchover between the two suit-circuit fans occur. The master caution light, warning and caution annunciator light assembly, and suit-circuit fan indicating lights will be monitored for switchover. Failure of second suit-circuit fan will be simulated to verify proper switching sequence. Failure indicating capabilities will be verified in both manual modes.

4-54. WASTE MANAGEMENT SYSTEM - EXTERNAL LEAKAGE CHECK

The purpose of this test will be to verify that the external leakage rate of the waste management system is within specified limits. This test will be conducted without the use of the space suit assembly.

A pressure controlled gaseous helium supply will be coupled to the Waste Management, and blanket pressure will be vented to the atmosphere. Gaseous helium will be supplied to the section and a selector valve within the section will be adjusted to permit the leak tracer gas to completely fill the section. Components and plumbing will be monitored for external leakage by utilizing a leak detector and monitoring pressure loss. At the completion of the test, the leak tracer gas will be vented and the section purged and conditioned to an evacuated state.

4-55. CABIN, INSTRUMENT, AND PANEL LIGHTING FUNCTIONAL TEST

The purpose for this test is to verify that the cabin instrument and panel lighting is operating satisfactorily. All lighting will be visually inspected and checked to verify that all lamps are illuminated and that they vary in intensity with actuation of the individual control, where applicable. Illumination will be accomplished either by switching or operating the test button.

4-56. FUEL CELL ASSEMBLY OPERATION IN THE ENVIRONMENTAL CHAMBER (MISSION SIMULATION)

This check is performed to demonstrate that the FCA performs efficiently and that the subsection's thermal characteristics under load meet flight requirements. During environmental chamber tests a gaseous reactant supply located outside the chamber will be used. The fuel cell assembly will be started prior to decompression and the gaseous distribution network will supply reactants and remove fuel cell assembly discharge fluids without allowing any to be released into the chamber. FCA start up will consist of removal of the inert pressurization gases, pressurization of the FCA to operational temperature. The FCA will be operationally tested by supplying power to all subsystems during the mission simulation. The amount of reactant used and the amount of power generated will be monitored to assure proper operation. The thermal characteristics of the FCA will be determined by monitoring the individual FCA internal and external temperature during chamber operation.

4-57. LEM/CSM - ECS INTERFACE TEST

The purpose of this check is to verify that the CSM-LEM interface is compatible with the LEM at altitude conditions.

After completion of LEM mission simulation tests in the altitude chamber, the LEM will be transposed to the other chamber after the Command/Service module has completed its altitude testing. Docking will be accomplished between the LEM overhead docking hatch and the Command/Service module in the vertical position. After docking, and with the crew in the Command/Service module, the chamber will be depressurized and a simulated mission test of the LEM ECS startup and post-rendezvous shutdown will be accomplished.

4-58. LEM/CSM RFI AND EMI COMPATIBILITY CHECK

The purpose of this check is to determine the effects of EMI from the Command/Service module on LEM System operation and to verify LEM and CSM rendezvous radar/transponder compatibility. The LEM/CSM EMI compatibility check will be performed by mounting the LEM in a test fixture and directing the radiating devices of the Command/Service module at the LEM so as to subject the LEM to the worst possible radiation front. All radiation-sensitive equipment in the LEM will be monitored for effects of this radiation. An LGS self-check will be performed during this check. The compatibility of the Rendezvous Radar and the Command/Service module will be checked by manually positioning the radar antenna to the approximate location of the Command module's transponder. Antenna search will be initiated, the acquisition and lock-on of the radar will be verified, and radar and transponder operation will be monitored. Radar tracking loops will be checked during the EMI compatability check by moving the Command/Service module and monitoring the radar tracking loops for effects of this movement. This check will be repeated for the Command/Service module radar and LEM transponder.

4-59. LANDING GEAR INSTALLATION TEST.

The purpose of this test is to install the landing gear on the descent stage of the LEM. This test will precede installation of the mated ascent and descent stages in the S-IVB adapter. The mated LEM will be in place on a support stand that will allow functional tests of the landing gear. The gear will be installed as directed, with the exception that live pyrotechnics for releasing the landing gear uplocks will be installed, but not armed.

4-60. MATE TEST, LEM TO S-IV-B ADAPTER

The purpose of this procedure is to mate the LEM with the S-IV-B adapter prior to their being stacked on the booster in the Vertical Assembly Building. The lower S-IVB adapter section will be placed on an adapter transporter in the integrated test area. The retracted landing gear will be installed on the descent stage. Pyrotechnic devices such as explosive bolts will be installed where required, but will not be armed until later in the checkout procedure.

4-61. VERTICAL ASSEMBLY BUILDING

4-62. MATING LEM TO LAUNCH VEHICLE

The purpose of these procedures is to secure the LEM to the launch vehicle. The LEM SIVB adapter assembly will be hoisted to its position on top of the SIVB and secured. All umbilical connections will be made from the Launch Umbilical Tower to the LEM.

4-63. FLIGHT CONTROL SYSTEM CHECKOUT

The purpose of this test will be to verify the satisfactory operation of the flight control system. The flight control system will be checked for operation, with emphasis placed upon the power turn-on sequence, high-low input power, Guidance and Navigation System operation, AGS Operation, rendezvous Radar RF checks, Automatic attitude control operation, attitude hold control, landing radar RF checks, manual flight control operation, descent control, and ascent control checks.

4-64. POWER DISTRIBUTION SECTION (PDS) CHECKOUT.

The purpose of this check is to verify the auto/manual operation of the PDS on external sources of power. The PDS will be checked by using operational tests detailed earlier in this section.

4-65. COMMUNICATIONS FUNCTIONAL CHECK.

The purpose of this check is to verify the overall operation of the Communications Subsystem in the integrated LEM. Functional testing will be accomplished on the communications, telemetry, and emergency communications facilities to ensure that they operate.

4-66. LAUNCH PAD.

4-67. FINAL CHECKOUT.

The mated stages of the LEM will be subjected to a final checkout of all systems to ensure proper operation of the LEM. This checkout will include operational and functional checks of the Environmental Control Subsystem, communications equipments, Flight Control Subsystem, LEM/Launch Complex EMI check, Rendezvous radar/transponder compatability check, Propulsion and Reaction Control Subsystems leakage tests, propulsion feed manifolds and thrust chamber assembly functional and leakage tests, reaction control system cold flow and injector response reaction control H_2 supply and blanket pressure tests, and a pyrotechnic power on-off voltage check.

4-68. SIMULATED COUNTDOWN

This procedure is performed to verify that the Power Generation Section can be serviced, started and operated within allowable time limits, and that the PGS is in acceptable condition for launch countdown. The cryogenic supply section tanks will be purged of inert gasses, filled with liquid reactants, cold-soaked, topped off, and sealed. The Fuel Cell Assembly will be purged, pressurized with O_2, H_2 and N_2 and brought to operating temperatures.

At the appropriate time, the descent tanks will be brought to operating pressure and the PGS will be switched to the internal reactant supply. The FCA's will then be connected to an electrical load. The PGS will be tested to determine that it can provide the electrical power required for selected portions of the mission while remaining within the operational parameters. At the completion of this test, the PGS will be shut down and returned to its onboard storage configuration.

4-69. Electrical Power Subsystem (PDS) Checkout. The objective of this test will be to verify the operation of the PDS under simulated countdown operations, and to verify the appropriate instrumentation. During countdown simulation of terminal the PDS will be switched over to internal fuel cell power, controls positioned to auto operation - normal mode, and bus voltages checked.

4-70. Fluid Servicing.

The purpose of this procedure will be to familiarize the launch pad personnel with the pre-launch countdown operations, and to load, drain, and purge the propulsion fluid systems. With the RCS installed in a mated LEM, on the launch pad, a simulated pre-launch checkout will be run: The fuel, and then the oxidizer, will be loaded. The helium tanks will be loaded, and the helium lines will be connected until the simulated close-out, at which time the helium tanks will be vented. Upon completion of the simulated countdown, the propellant tanks will be drained, cleaned, and flushed, using a nitrogen purge, a de-ionized water flush, and a neutralizer flush. Nitrogen at blanket pressure will be applied to the bladders upon completion of the cleaning process.

4-71. COUNTDOWN.

4-72. EPS-PGS Checkout. The objective of this procedure will be to place the PGS in its launch configuration. This test will involve the complete checkout and servicing of the PGS, including switchover to internal power for launch. All essential PGS parameters will be monitored by ACE, or onboard instrumentation, or both to verify that the section is performing properly. The sequence of events will include: CSS fill-condition and top-off, FCS startup, and PGS operation.

4-73. Communications Functional Tests. The purpose of this test will be to perform a final communication subsystem check prior to attaining final launch configuration. A complete checkout of all communications, Telemetry, TV and emergency communication facilities will be accomplished.

4-74. Propellant Loading. The purpose of this procedure will be to fill the ascent and descent propellant tanks with the required quantities of conditioned propellants. Both the ascent and descent fuel tanks will be filled simultaneously by the same method. After the fueling is completed, both the ascent and descent oxidizer tanks will be filled simultaneously.

4-75. ECS Countdown. The objective of this procedure will be to prepare the ECS for launch. The Heat Transport Section will provide cooling to the electronic cold plates within approximately T-8 hours using GSE-circulated water-glycol. At T-8 hours the Heat Transport Section recirculation assembly will be energized, GSE uncoupled, and Freon supplied to the Freon boilers, which will maintain temperature control. The LEM hatches will be closed and latched and the LEM cabin pressurized to approximately 5.6 psia above ambient pressure.

4-76. Pyrotechnic Installation. To mechanically install the pyrotechnic devices a bridge wire continuity check and an electrical isolation test will be performed on all EED's, using portable pyrotechnic continuity and electrical isolation testers. The pyrotechnic devices will then be brought to the LEM and installed in their proper places. The electrical connections will not be made at this time; instead shorting connectors will be on the EED's until electrical connections are made. Positive identification as to location of each device installed in LEM will be performed.

4-77. Removal of Non-Flight Hardware. All non-flight items will be removed from the LEM and the S-IV-B adapter prior to firing. This includes all servicing lines, ACE carry-on and associated equipment, work platforms and other GSE. All servicing parts in the S-IVB adapter will be closed as their associated lines are removed. After removal of non-flight hardware from the interior of the adapter, the access hatches will be closed and secured. Shortly before launch approximately 30 seconds), the umbilicals from the LUT to the adapter will be pulled by remote control.

SECTION V

GROUND SUPPORT EQUIPMENT

5-1. GENERAL.

The LEM Ground Support Equipment (GSE) consists of the LEM Carry-On Equipment, Spacecraft and Servicing Equipment Control and Checkout Equipment, Servicing Equipment, Condition and Equipment Simulators, Handling and Transportation Equipment and Workstands, and Bench Maintenance Equipment.

5-2. LEM CARRY-ON EQUIPMENT.

The LEM Carry-On Equipment is a spacecraft-associated extension of the Acceptance Checkout Equipment - Spacecraft (ACE-S/C) Ground Station. ACE-S/C is a general-purpose spacecraft checkout system used to perform computer-controlled or manually controlled testing of the LEM system and subsystems. The ACE-S/C Ground Station control room, digital computers, data acquisition and decommutation equipment, and auxiliary support equipment are in the Operations and Checkout Building. The LEM Carry-On Equipment adapts the ACE-S/C Ground Station to the LEM subsystems and servicing equipment. The Carry-On Equipment used with the LEM subsystems is, for the most part, carried on board the LEM when ACE-S/C-controlled prelaunch tests are to be performed.

The LEM Carry-On Equipment comprises two ACE-S/C Up-Links, the Communications Subsystem Checkout Equipment, the Radar Checkout Equipment, and an ACE-S/C Down-Link. Digital test commands generated by the ACE-S/C Ground Station and addressed to the LEM subsystems are conditioned and routed to their proper input points by one ACE-S/C Up-Link. The second ACE-S/C Up-Link performs the same function for test commands addressed to the LEM servicing equipments. Special radio frequency tests of the LEM Communications Subsystem and the two radars are performed by the Communications Subsystem Checkout Equipment and the Radar Checkout Equipment. The ACE-S/C Down-Link monitors, samples, conditions, and interleaves LEM subsystem performance data with LEM telemetry data and with LEM servicing equipment response data; the resultant data train is transmitted to the ACE-S/C Ground Station.

5-3. ACE-S/C UP-LINK.

The ACE-S/C Up-Link performs the function of a Digital Test Command System (DTCS) and a Stimuli Generator. Together, these can generate up to 2,048 discrete or analog stimuli, or 128 single-point differential output analog signals, or various combinations thereof, and route these stimuli to the ACE-S/C Ground Station-addressed onboard equipment or servicing equipment input. Upon receipt of an ACE-S/C Ground Station command, the DTCS transmits a "check status reply" to the ACE-S/C Ground Station for validity analysis, automatically providing continuous self-check information to assure the accuracy of the checkout data processed by the ACE-S/C Up-Link.

The DTCS used in the LEM comprises the following assemblies:

a. Receiver-decoder. This assembly receives ACE-S/C Ground Station test commands at rates up to one million bits per second and routes test commands to a baseplate unit group or the guidance and navigation (G&N) computer buffer unit.

b. G & N computer buffer unit. This assembly processes, checks, and routes test commands received from the receiver-decoder and addressed to the LEM G & N Subsystem's Guidance Computer.

c. Baseplate unit. Up to eight baseplate units can be used to form one baseplate unit group; up to four baseplate unit groups can be used simultaneously. All receiver-decoder outputs not intended for the LEM Guidance Computer are addressed to one of the baseplate unit groups and to one baseplate unit within that group. The addressed baseplate unit selects one of four modules at its output. The four output modules can be any combination of the following three modules (d, e, and f). The combination of modules used depends upon the type of input required by the LEM subsystem undergoing test.

d. Conventional relay module. Upon ACE-S/C Ground Station command applied to it by a baseplate unit, the conventional relay module controls the application of stimuli to the LEM subsystems.

e. Latching relay module. This assembly performs the same function as the conventional relay module. The latching relay module relays change state only upon DTCS command, not upon power application or removal. The latching relay module is used in situations where loss of power to a conventional relay that controls a critical signal or bias would cause the relay to drop out and subsequently damage the subsystem undergoing test.

f. Digital-to-analog converter module. This assembly converts its binary output from the baseplate unit into analog signals for application to the LEM subsystems.

The Stimuli Generator is a special stimuli unit that provides all necessary stimuli not supplied by the DTCS.

The ACE-S/C Up-Link that is used with the LEM servicing equipment includes all the assemblies previously described, except the G & N computer buffer unit.

5-4. COMMUNICATIONS SUBSYSTEM CHECKOUT EQUIPMENT.

The Communications Subsystem Checkout Equipment enables ACE-S/C-controlled automatic testing of the LEM Communications Subsystem. The capability of ACE-S/C to accurately generate and measure radio frequency signals under computer control enables comprehansive tests that: (1) exercise the entire Communications Subsystem in all modes of operation, and (2) test the capability of the S-band receiver to track varying frequencies, such as are the effects of doppler shifts. The Communications Subsystem Checkout Equipment accepts DTCS commands to simulate an earth communications station that generates LEM receiver inputs and monitors LEM transmitter outputs. Transmitter frequency and power outputs, and receiver sensitivity and bandwidth are monitored, as well as such non-RF parameters as AGC voltage, phase-lock error voltage, audio frequency, ranging voltage inputs to the transmitters, and power supply voltages. Selection of the various modes of operation of the Communications Subsystem during testing will be made by an astronaut or a test conductor at the LEM controls.

5-5. RADAR CHECKOUT EQUIPMENT.

The Radar Checkout Equipment is used to automatically test the Landing Radar and rendezvous radar/transponder of the LEM Navigation and Guidance Subsystem. Using command stimuli applied to it by the DTCS, the Radar Checkout Equipment causes the radars to be exercised through their performance capabilities. Radar parameters that are automatically tested include: acquisition threshold, acquisition time, acquisition scan limits, track threshold, range tracking, velocity tracking, angle tracking, and RF output power and frequency. These parameters are monitored by measuring equipment with RF-adaptive capabilities, and the resulting response data are applied to signal conditioners, and then to the Carry-On PCM system of the ACE-S/C Down-Link.

5-6. ACE-S/C DOWN-LINK.

The LEM subsystem and servicing equipment responses to the test stimuli (applied by the DTCS and the special stimuli generator) are monitored by the ACE-S/C Down-Link, which acts as a Digital Test Monitor System (DTMS). The DTMS performs three major functions; that of a Pulse Code Modulation (PCM) Response Section, a Servicing Equipment - ACE-S/C Adapter, and a Data Interleaver. Together, these process all LEM checkout data into a form acceptable to the ACE-S/C Ground Station.

The PCM Response Section comprises the following assemblies:

a. Digital signal conditioning and multiplexing unit. Using timing signals from the carry-on PCM system (f, below), this assembly conditions and time multiplexes up to 318 event (on-off) signals from the LEM subsystems into an eight-line parallel format for transmission to the carry-on PCM system.

b. Analog signal conditioning and sampling unit. Using carry-on PCM system timing signals, this assembly samples up to 250 analog signals from the LEM subsystems at a rate of one sample per second and up to 50 analog signals at a rate of ten samples per second. These are transferred to the carry-on PCM system in a pulse-amplitude-modulated (PAM), non-return-to-zero (NRZ) format.

c. Special signal conditioning unit. This assembly performs much the same function as the analog signal conditioning and sampling unit, and is used when nonstandard analog outputs of the LEM subsystems make special analog signal conditioning circuitry necessary.

d. High-sampling-rate signal conditioning unit. Using carry-on PCM timing signals, this assembly accepts 50 unconditioned and 20 conditioned signals from the LEM subsystems. Of the 50 unconditioned signals, 20 are conditioned and applied to the carry-on PCM system. The remaining 30 unconditioned signals are conditioned and subcommutated with the 20 conditioned input signals. These are applied to the carry-on PCM system in a PAM format.

e. G & N signal conditioning and switching matrix unit. This assembly receives 39 unconditioned analog signals and 11 conditioned analog signals from the LEM G & N Subsystem. Of the 39 unconditioned analog signals, 32 are routed to this assembly's switching matrix unit, where one of the signals is selected for sampling by the carry-on PCM system at a rate of 400 samples per second. All of the input signals are applied to this assembly's 50 channel subcommutator, where they are subcommutated into a PAM data train and routed to the carry-on PCM system.

f. Carry-On PCM system. The inputs to the carry-on PCM system are timing signals from the data interleaving system, and 128 channels of analog data, PAM-NRZ data, and digital data from the signal conditioning units described previously. The carry-on PCM system supplies timing signals to these signal conditioning units, and converts and interleaves the input data from the 128 data channels into a PCM-NRZ format for transfer to the data interleaving system.

The Servicing Equipment - ACE-S/C Adapter is an assembly that receives timing signals from the data interleaving system and up to 1,000 event signals and 200 analog signals from the LEM servicing equipments. The servicing equipment data are multiplexed into a PCM-NRZ data train and transferred to the data interleaving system.

The Data Interleaver function is performed by the data interleaving system assembly, which interleaves LEM subsystem performance data from the carry-on PCM system, data supplied by the LEM telemetry, and data from the servicing equipment - ACE-S/C adapter assembly. Operation of the data interleaving system assembly is controlled by command signals from the DTCS and timing signals from the LEM telemetry. The interleaved LEM subsystem, telemetry, and servicing equipment data is transmitted in a return-to-zero, bi-phase, PCM format to the ACE-S/C Ground Station. The data interleaving system assembly also sends non-interleaved telemetry data to the ACE-S/C Ground Station.

5-7. SPACECRAFT AND SERVICING EQUIPMENT CONTROL AND CHECKOUT EQUIPMENT.

5-8. TEST CONDUCTOR CONTROL STATION.

The Test Conductor Control Station is a central control console that enables two test overseers to supervise all operations concerned with Propulsion Subsystem and Reaction Control Subsystem testing. The station has interfaces with the: (1) Propellant Control Station, (2) Pneumatics Control Station, (3) RCS Stand Control Station, (4) Engine Programmer, (5) Engine Firing Control Station, (6) RCS Programmer, and (7) RCS Firing Control Station. These interfaces enable: (1) displays that provide status indications of the overall test and of the previously mentioned equipments, (2) power interlock controls over these equipments, and (3) the necessary control functions to place a test in a "hold" or "abort" status. The Test Conductor Control Station includes a master intercommunication station that enables the test overseers to converse with the operators of the previously mentioned equipments and to tie into the facility public address system.

5-9. PROPELLANT CONTROL STATION.

The Propellant Control Stations are remote control consoles used in the Static Test Area for the control of: (1) the conditioning and transfer of propellants (fuel and oxidizer) from ready storage to the test stand, (2) distribution of propellants throughout the test stand to and from the spacecraft interfaces, (3) loading of the LEM Propulsion Subsystem propellant tanks, and (4) management of toxic vapors and waste propellants. One Propellant Control Station controls the transfer of fuel and a second controls the transfer of oxidizer. In accomplishing these control functions, the Propellant Control Stations will remotely control: (1) a Fuel Ready Storage Unit, (2) an Oxidizer Ready Storage Unit, (3) a Fuel Transfer and Conditioning Unit, (4) an Oxidizer Transfer and Conditioning Unit, (5) the Fuel Loading Control Assembly, (6) the Oxidizer Loading Control Assembly, (7) all associated test stand fluid distribution systems, (8) a Fuel Toxic Vapor Disposal Unit, (9) an Oxidizer Toxic Vapor Disposal Unit, (10) the facility fuel and oxidizer dump tanks, and (11) all valves in the facility fuel and oxidizer lines. The switching logic of each Propellant Control Station, which has start-safe, fail-safe, and emergency stop capabilities, is such that inadvertent activation of a mode that would be detrimental to the equipment or dangerous to personnel is impossible. Interfaces with the Test Conductor Control Station enable the Propellant Control Station to apply discrete signals that are used indicate its own status and that of the loading cycle to the test overseers, and to receive "hold" or "abort" commands.

5-10. PNEUMATICS CONTROL STATION

The Pneumatics Control Station is a remote control console used in the Static Test Area for the control of: (1) the conditioning and transfer of gaseous helium from ready storage to the test stand, (2) the distribution of gaseous helium throughout the test stand to and from the spacecraft Propulsion Subsystem pneumatics interfaces, (3) the loading of the LEM Propulsion Subsystem pneumatics section, and (4) the transfer of gaseous nitrogen from ready storage to the test stand. In accomplishing this the Penumatics Control Station will remotely control: (1) a Helium Transfer and Conditioning Unit, (2) a Helium Pressure Distribution Unit, (3) all test facility electrically controlled valves within the helium and nitrogen lines, (4) the test stand nitrogen vacuum breaker, (5) emergency shutdown of the Helium Booster Cart, and (6) all associated test stand fluid distribution systems. The switching logic which has start-safe, fail-safe, and emergency stop capabilities is such that inadvertent activation of a mode that would be detrimental to the equipment or dangerous to personnel is impossible. Interfaces with the Test Conductor Control Station enable the Pneumatics Control Station to apply discrete signals that are used to indicate its own status and that of the loading cycle to the test overseers, and to receive "hold" or "abort" commands.

5-11. RCS STAND CONTROL STATION.

The RCS Stand Control Station combines the functions of the Propellant Control Station and the Pneumatics Control Station (which are for the Propulsion Subsystem) to control the fuel, oxidizer, and pneumatics transfer operations associated with the LEM Reaction Control Subsystem in the Static Test Area. The switching logic incorporates the same safety factors, and similar interfaces with the Test Conductor Control Station are included. The RCS Stand Control Station will remotely control: (1) RCS Fuel Servicing Unit, (2) RCS Oxidizer Servicing Unit, (3) RCS Fuel Transfer Control Unit, (4) RCS Oxidizer Transfer Control Unit, and (5) all test stand RCS vent, bleed, and drain valves.

5-12. PROPELLANT LOADING CONTROL ASSEMBLY CONTROLLER.

The Propellant Loading Control Assembly Controller enables remote manual control of the Fuel Loading Control Assembly or the Oxidizer Loading Control Assembly when these are used for launch operations, such as at the Mobile Arming Tower. (At the Static Test Area, Propellant Control Stations remotely control the Fuel and Oxidizer Loading Control Assemblies.) The capabilities of the Propellant Loading Control Assembly Controller include: (1) controlling the Fuel or Oxidizer Loading Control Assembly through all modes of operation, (2) monitoring fuel or oxidizer pressures, temperatures, flow rates, and quantity loaded, (3) commanding the Loading Control Assembly to a safe condition should an emergency develop, and (4) supporting maintenance and calibration of either Loading Control Assembly.

5-13. HELIUM DISTRIBUTION UNIT CONTROLLER.

The Helium Distribution Unit Controller enables remote manual control of the Helium Pressure Distribution Unit when it is used for launch operations, such as at the Mobile Arming Tower. (At the Static Test Area, the Pneumatics Control Station remotely controls the Helium Pressure Distribution Unit.) The Helium Distribution Unit Controller's capabilities include: (1) controlling the Helium Pressure Distribution Unit through all modes of operation, (2) monitoring ascent stage, descent stage, and RCS helium pressure from 0 to 3,500 psig, (3) monitoring helium temperature within a range of plus or minus 200 degrees, and (4) emergency venting of the Helium Pressure Distribution Unit.

5-14. PROPULSION SUBSYSTEM CHECKOUT GROUP.

The Propulsion Subsystem Checkout Group is used along with the Propulsion Subsystem Checkout Set to perform prefiring and postfiring static tests and dynamic nonfiring checks of the LEM Propulsion Subsystem to verify subsystem integrity for flight. Electrical continuity checks, resistance checks, and insulation resistance checks are automatically scanned between preset "no" and "no-go" limits, providing automatic end-to-end electrical testing. The Propulsion Subsystem Checkout Group will supply primary power and all required combinations of input stimuli to the subsystem under test. The switching logic, which has start-safe, fail-safe, and emergency stop capabilities, is such that inadvertent activation of a mode that would be detrimental to the equipment under test or dangerous to personnel is impossible. Readout devices include analog and digital displays and recorders.

5-15. PROPULSION SUBSYSTEM CHECKOUT SET.

The Propulsion Subsystem Checkout Set consists of mobile units that are grouped into three main categories: pressurization checkout, feed system checkout, and engine checkout. The units of the checkout set are connected to the test ports of the subsystem. Pneumatic stimuli are applied through the test ports and return passages are monitored. The checkout proceeds from point-to-point until complete subsystem is checked out. All units are manually and locally operated except where safety procedures dictate remote control.

5-16. RCS CHECKOUT GROUP.

The RCS Checkout Group is used with the RCS Checkout Cart to perform prefiring and postfiring static tests and dynamic nonfiring checks of the LEM Reaction Control Subsystem, to verify subsystem integrity for flight. Various electrical tests of the RCS solenoid valves, thrust chamber assemblies, and transducers are also enabled, providing end-to-end electrical verifications. The RCS Checkout Group will supply primary power and all required combinations of input stimuli to the RCS, and will monitor and display appropriate outputs and test points. The switching logic, which has start-safe, fail-safe, and emergency stop capabilities, is such that inadvertent activation of a mode that would be detrimental to the equipment under test or dangerous to personnel is impossible.

5-17. RCS CHECKOUT CART.

The RCS Checkout Cart is a mobile test facility used for performance and leak testing of the Reaction Control Subsystem or components of the subsystem individually or while assembled in the RCS. Gaseous nitrogen or helium is supplied to the RCS or components from the cart at various controlled pressures and/or flow conditions. The cart also provides facilities for vacuum pumping of the pressure regulator reference ports and can be used in conjunction with a mass spectrometer leak detector for individual leak checking of RCS components. A four-wheel platform, with towing facilities, supports the cart and provides ease of mobility. The cart is towed into position and, after being connected to the RCS or an RCS Components, is operated by its control panels.

5-18. ENGINE PROGRAMMER.

The Engine Programmer is used to control the Engine Firing Control Station automatically, enabling either the ascent stage propulsion engine or the descent stage propulsion engine to be fired in a simulated mission profile, or in special profiles incorporating simulated equipment failures. The mission profiles are preprogrammed on digital recording tape which is used by the Engine Programmer in generating Propulsion Subsystem command stimuli. The Engine Programmer has interfaces with the Test Conductor Control Station, and with the RCS Programmer. The interfaces with the Test Conductor Control Station enable the Engine Programmer to apply discrete signals that are used to indicate its own status and that of the test to the test overseers, and to receive "hold" or "abort" commands. Interfaces with the RCS Programmer enable the Engine Programmer to operate in synchronization with the RCS Programmer. This capability enables simultaneous testing of the LEM ascent stage propulsion engine and the RCS engines: the two Programmers and their Firing Control Stations automatically fire the ascent engine and the RCS engines in simulated ascent, rendezvous, and docking mission profiles.

5-19. ENGINE FIRING CONTROL STATION.

The Engine Firing Control Station is used to fire the LEM ascent stage and descent stage propulsion engines during sea level or simulated altitude static tests. When used alone, it provides manual control of engine firing, and when used with the Engine Programmer, it provides automatic control of engine firing in simulated mission profiles. Manual control is enabled by two sets of plug-in control panels, one set for the ascent engine and one set for the descent engine. The switching logic, which has start-safe, fail-safe, and emergency stop capabilities, is such that inadvertent activation of a mode that would be detrimental to the equipment under test or dangerous to personnel is impossible. The Engine Firing Control Station's capabilities include: (1) the necessary control functions to enable a comprehensive prefiring checkout of the propulsion engine when mounted in the test stand, (2) firing of the engine either directly or through the Stabilization and Control Subsystem, (3) flexible control of the engine firing time, (4) monitoring and displaying of test parameters to enable the operator to determine overall test status, and (5) emergency shutdown of the engine under test. Interfaces with the Test Conductor Control Station enable the Engine Firing Control Station to apply discrete signals that are used to indicate its own status and that of the test to the test overseers, and to receive "hold" or "abort" commands.

5-20. RCS PROGRAMMER.

The RCS Programmer is used to control the RCS Firing Control Station automatically, enabling the RCS engines to be fired in a simulated mission profile, or in special profiles incorporating simulated equipment failures. Automatic tape control is used by the RSC Programmer in generating command stimuli that cause translation and rotation of the spacecraft to be simulated. The RCS Programmer has interfaces with the Test Conductor Control Station and with the Engine Programmer. The interfaces with the Test Conductor Control Station enable the RCS Programmer to apply discrete signals that are used to indicate its own status and that of the test to the test overseers, and to receive "hold" or "abort" commands. Interfaces with the Engine Programmer enable the RCS Programmer to operate in sychronization with the Engine Programmer. This capability enables simultaneous testing of the RCS engines and the ascent stage propulsion engine: the two Programmers and their Firing Control Stations automatically fire the RCS engines and the ascent engine in simulated ascent, rendezvous, and docking mission profiles.

5-21. RCS FIRING CONTROL STATION.

The RCS Firing Control Station is used to control the operation and firing of the Reaction Control Subsystem in static tests that allow evaluation of the RCS pressurization section, propellant sections, and thrust characteristics, and also enable adjustment of the various RCS parameters during a test run. When used alone, the RCS Firing Control Station provides manual or semiautomatic control of the RCS, and when used with the RCS Programmer, automatic firing in simulated mission profiles is enabled. The switching logic, which has start-safe, fail-safe, and emergency stop capabilities, is such that inadvertent activation of a mode that would be detrimental to the equipment under test or dangerous to personnel is impossible. The RCS Firing Control Station's capabilities include: (1) the generation of engine firing control signals that are continuous, or that have a pulse repetition rate from 0.00125 to 25 cps and a pulse width from 10 milliseconds to 800 seconds, (2) selection of a given RCS engine without using or simulating the jet select logic of the Stabilization and Control Subsystem (SCS), (3) firing the RCS engines through the SCS, (4) the necessary steering logic to simulate translation, attitude, and attitude hold maneuvers by simulating either singly or in any combination, thruster pairs, thruster couples, and thruster reversing couples, (5) simulation of malfunctions by the closing and opening of the RCS valves, or by applying discrete signals to the SCS, and (6) displaying the status of the RCS valves and 16 engines. Interfaces with the Test Conductor Control Station enable the RCS Firing Control Station to apply discrete signals that are used to indicate its own status and that of the test to the test overseers, and to receive "hold" or "abort" commands.

5-22. REACTION CONTROL AND PROPULSION SUBSYSTEM COMPONENT TEST STATION.

The Reaction Control and Propulsion Subsystem Component Test Station is used, along with the Helium Components Test Stand, the Referee Fuel Components Test Bench, and the Referee Oxidizer Components Test Bench to perform incoming inspection, acceptance testing, and maintenance of the various Propulsion Subsystem and RCS components to verify the integrity of the component for flight. This test station primarily enables control, testing, and display of the electrical parameters of the component under test. Pneumatic and fluid parameters are displayed on the corresponding component test stand or bench. The switching logic, which has start-safe, fail-safe, and emergency stop capabilities, is such that inadvertent actication of a mode that would be detrimental to the equipment or dangerous to personnel is impossible.

5-23. RCS PROPELLANT QUANTITY GAGING SYSTEM TEST SET.

The RCS Propellant Quantity Gaging System Test Set (PQGS) is used to test and calibrate the LEM RCS propellant (fuel and oxidizer) quantity gaging system. It is a portable test set that performs three major functions: (1) verify that the PQGS of the LEM is operating properly, (2) provide accurate readouts to enable calibration of the PQGS, and (3) isolate malfunctions to any one of the four onboard quantity sensor assemblies, or to the PQGS control unit. The PQGS Test Set has two major modes of operation and a self-test mode. The quantity sensor mode of operation is used to check the output of each quantity sensor assembly. The control unit mode enables monitoring and displaying of either both fuel quantities, or both oxidizer quantities, and obtains these quantities directly from the PQGS control unit, as do the astronaut displays. A simple comparison between the test set readouts and a known quantity of propellant loaded into the LEM tanks will indicate whether the PQGS is operating properly. This mode can be used to verify proper operation of the PQGS control unit alone by simulating the quantity sensor inputs to the control unit. The operator then checks that the control unit quantity outputs correspond to the simulated inputs. Calibration of the PQGS control unit is accomplished in this mode by taking quantity readings when the LEM tanks are empty. The empty tank readings are used to determine the settings for the PQGS control unit calibration devices.

5-24. PYROTECHNIC INITIATOR TEST SET.

The Pyrotechnic Initiator Test Set is used to verify the prefiring integrity of squib valve igniter bridgewires. Testing is nondestructive and completely safe for both the operator and the equipment. The Pyrotechnic Initiator Test Set will measure and directly indicate the resistance of the squib valve bridgewire while maintaining the test current at a level well below the firing current of a single squib.

5-25. POWER GENERATION SECTION SPACECRAFT AND SERVICING REMOTE CONTROL CONSOLE.

This console is used to perform complete static and dynamic testing of the power generation section of the Electrical Power Subsystem, and is a central control console for the various supporting and servicing equipments that are used for EPS testing. It enables control and monitoring the following equipment: (1) Gaseous and Cryogenic Fluid Distribution Network Control Console, (2) Power Generation Section Heater Power Supply, (3) Load Simulator, (4) Gaseous Hydrogen Supply Unit, (5) Gaseou Oxygen Supply Unit, (6) Referee Cryogenic Fluid Supply Unit, (7) Gaseous Nitrogen and/or Helium Supply Unit, (8) Hydrogen Transfer and Monitoring Unit, (9) Oxygen Transfer and Monitoring Unit, (10) Liquid Hydrogen Mobile Storage Unit, and (11) Liquic Oxygen Mobile Storage Unit. The Power Generation Section Spacecraft and Servicing Remote Control Console's capabilities include: (1) display of the status of the overall test and of the previously mentioned equipments, (2) power interlock controls over the previously mentioned equipments, and (3) the necessary control functions to place a test in a "hold" or "abort" status. A master intercommunication station enables the operators to converse with the operators of any of the previously mentioned equipments. This console has interfaces with ACE-S/C.

5-26. GASEOUS AND CRYOGENIC FLUID DISTRIBUTION NETWORK CONTROL CONSOLE.

This console is a central control station for the test area facility lines, valves, and valving interfaces that are used to transfer gaseous and liquid hydrogen, gaseous and liquid oxygen, gaseous helium, and gaseous nitrogen. It is used to verify that the facility equipment is operating properly, and to control purging, cold-soaking, filling, draining, and detanking of the power generation section of the Electrical Power Subsystem. Operation of this console can be local or remote; its interfaces enable it to operate in conjunction with the Power Generation Section Spacecraft and Servicing Remote Control Console or with ACE-S/C.

5-27. POWER GENERATION SECTION HEATER POWER SUPPLY.

The Power Generation Section Header Power Supply is used to control and monitor the application of ground power to the spacecraft fuel cell and cryogenic storage system heaters during power generation section testing and other prelaunch operations. If fuel cell heater power is not required, this power supply will separately control and monitor the application of power to both the ascent stage and descent stage cryogenic storage system heaters. Current crossover limiting prevents heater overload. Interfaces enable this unit to operate in conjunction with the Power Generation Section Spacecraft and Servicing Remote Control Console or with ACE-S/C.

5-28. VEHICLE POWER SUPPLY UNIT.

The Vehicle Power Supply Unit is used to supply prime d-c power to the LEM vehicle under test when the spacecraft fuel cell assemblies are not used. It will supply three separate output voltages adjustable from 15 to 36 volts dc, each with a peak power rating of 1,300 watts, to the power distribution section of the Electrical Power Subsystem. Automatic shutdown devices are included as protection against overload and reverse current conditions.

5-29. RENDEZVOUS RADAR/TRANSPONDER COMPATIBILITY CHECKOUT EQUIPMENT.

The Rendezvous Radar/Transponder Compatibility Checkout Equipment is used to verify the ability of the LEM Rendezvous Radar and the Command Module Transponder to lock onto each other. If a fault exists, it will be isolated to either the Rendezvous Radar or the Transponder. In performing compatibility checks, this equipment will test the following parameters: range output, range rate output, azimuth and elevation angles, angle rates, acquisition threshold, acquisition time, frequency, spectrum, output power, pulse repetition frequency, receiver AGC voltages, and crystal currents.

5-30. GASEOUS COMPONENTS TEST BENCH.

The Gaseous Components Test Bench is a test stand that consists of a pump, blowers, valving and plumbing, and controls and instrumentation. It is designed for use in a Class II clean room and is used to test all components of the atmosphere revitalization and oxygen supply sections. This unit is capable of measuring pressure-drop across components, testing flow rate, proof-testing of components, testing of electrical component efficienty, and determining internal and external leakage.

5-31. ATMOSPHERE REVITALIZATION CALIBRATION TEST SET.

The Atmosphere Revitalization Calibration Test Set is a unit that imposes the metabolic loading of 0 to 2 men upon the atmosphere revitalization section of the Environmental Control Subsystem during subsystem check out. This unit is an item of common usage GSE.

5-32. WATER COMPONENTS TEST STAND.

The Water Components Test Stand is a test bench that contains a water pump and reservoir, a nitrogen purge section, a vacuum pump, a sink, a handpump, and the necessary instrumentation and connections. This test stand is used in a Class II clean room. An auxiliary unit accompanies this test stand and contains all the equipment requiring frequent maintenance and is located outside the clean room. The test stand is used to supply a component under test with operational inputs; to pressurize a component for proof-test; to perform internal and external leakage tests; to determine the amount of water in the tanks; to perform functional tests of all pressure, differential pressure, fluid level, and valve position indicator sensors; and to purge and evacuate the water components.

5-33. WATER-GLYCOL COMPONENTS TEST STAND.

The Water-Glycol Components Test Stand is a test bench that contains a water-glycol pump and reservoir; a water flush pump and reservoir; a nitrogen purge system; a vacuum pump; a temperature control unit; a sink; a handpump; and the necessary instrumentation, plumbing and connections. This test stand is used in a Class II clean room. An auxiliary unit accompanies this test stand and contains all the equipment requiring frequent maintenance and is located outside the clean room. The test stand is used to supply components under test with operational inputs of water-glycol; to pressurize components for proof-pressure tests; to perform internal and external leakage tests; to perform functional tests of all pressure, differential pressure, speed, fluid level and valve position sensors; to test electrical efficiency of pump; and to flush, purge, and evacuate the water-glycol components.

5-34. CABIN LEAKAGE TEST UNIT.

The Cabin Leakage Test Unit is a portable unit containing valves, flowmeters, a pressure regulator, plumbing, and pressure gages. This unit is used to measure leakage of the LEM cabin and to purge the LEM cabin with gaseous oxygen.

5-35. HELIUM COMPONENTS TEST STAND.

The Helium Components Test Stand is a structural steel stand with sliding steel doors, explosion-resistant glass viewing ports, test chamber, and bullet-resistant glass viewing ports, test chamber, and bullet-resistant mesh (located in the vent duct on top of the test chamber). Also included are the necessary controls and instrumentation. This unit is to be used in a sheltered location where temperature, humidity, and cleanliness are controlled. This test stand is used to manually control and measure the flow and pressure through the Propulsion and Reaction Control Subsystem helium pressure system components, and to proof-pressure test the components. The flow and pressure tests are conducted in the explosion-proof, bullet-resistant test chamber. An automatic safety circuit is provided to prevent the operator from opening the chamber doors during the pressurization tests. The Reaction control and Propulsion Subsystem Component Test Station remotely controls the operation of components tested on this unit.

5-36. REFEREE FUEL COMPONENTS TEST BENCH.

The Referee Fuel Components Test Bench is a test stand that contains a deionized water flow system, a static pressure system, and a control panel with the necessary controls and instrumentation. To simulate the fuel system flow characteristics, this unit provides manual control of flow and pressure, and a static pressure for proof-pressure testing of the fuel system components. The Reaction Control and Propulsion Subsystem Component Test Station remotely controls the operation of components tested on this unit.

5-37. REFEREE OXIDIZER COMPONENTS TEST BENCH.

The Referee Oxidizer Components Test Bench is a test stand that contains a freon flow system, a static pressure system, and a control panel with the necessary controls and instrumentation. This unit provides manual control of flow and pressure to simulate the oxidizer system flow characteristics, and a static pressure for proof testing of the oxidizer system components. The Reaction Control and Propulsion Subsystem Component Test Station remotely controls the operation of components tested on this unit.

5-38. HALOGEN LEAK DETECTOR.

The Halogen Leak Detector is a portable unit that contains a sensing probe which includes an indicating meter, sensitivity knob, and a zero-set control. This unit is used to check out a propulsion feed system component that has been pressurized with a tracer gas (freon). By passing the sensing probe over the component, this unit provides both a visual and audio signal, if a leak is present.

5-39. CRYOGENIC AND GASEOUS OXYGEN COMPONENTS TEST BENCH.

The Cryogenic and Gaseous Oxygen Components Test Bench is a test stand that contains cryogenic storage tanks, cryopiping, and the necessary controls and instrumentation. It is used to control and monitor the flow of liquid oxygen, gaseous oxygen, and supercritical oxygen through adapters for component checkout.

5-40. CRYOGENIC AND GASEOUS HYDROGEN COMPONENTS TEST BENCH.

The Cryogenic and Gaseous Hydrogen Components Test Bench is a test stand that contains cryogenic storage tanks, cryopiping, and the necessary controls and instrumentation for bench testing of the Electrical Power Subsystem hydrogen components. It is used to control and monitor the flow of liquid hydrogen, gaseous hydrogen, and supercritical hydrogen through adapters for component checkout.

5-41. SERVICING EQUIPMENT.

5-42. GASEOUS OXYGEN SUPPLY UNIT.

The Gaseous Oxygen Supply Unit is a mobile unit that contains a pressure regulator, valves, plumbing, a flowmeter, pressure gages, and a vacuum pump. It is used to evacuate the oxygen supply section of the Environmental Control Subsystem and to supply gaseous oxygen from the facility to charge the gaseous oxygen accumulator of the oxygen supply section. It also supplies gaseous oxygen to the ECS when the Electrical Power Subsystem supercritical oxygen tanks are not available.

5-43. WATER-GLYCOL TRIM CONTROL UNIT.

The Water-Glycol Trim Control Unit is a portable unit located in a sheltered area on the Mobile Arming Tower and contains a heat exchanger, a pump, valves, a temperature controller, relief valves, plumbing, wiring, a heater, and a reservoir. This unit is used to fill the heat transport loops with a water-glycol mixture, circulate the water-glycol mixture, and either add or remove heat from the mixture. The water-glycol mixture is supplied to this unit from the Water-Glycol Service Unit.

5-44. RCS FUEL TRANSFER CONTROL UNIT.

The RCS Fuel Transfer Control Unit is a four-wheel cart that contains fuel tanks, a weighing system, a control and instrumentation panel, filters, valves, and LEM and launch tower hardline adapters and hoses. When the unit is connected to the fuel hardlines and the Reaction Control Subsystem fill connections, it measures the weight of the fuel supplied, filters the fuel, and monitors the fuel temperature and pressure. At the Static Test Area, this unit is remotely controlled by the RCS Stand Control Station.

5-45. FUEL FLUSH AND PURGE CART.

The Fuel Flush and Purge Cart is a four-wheel cart that uses self-contained deionized water, hydroxyacetic acid solution, methyl alcohol, and gaseous nitrogen to provide cleaning and decontamination of the Reaction Control Subsystem tanks, systems, and components which have been exposed to the fuel (unsymmetrical dimethylhydrazine).

5-46. RCS OXIDIZER TRANSFER CONTROL UNIT.

The RCS Oxidizer Transfer Control Unit is a four-wheel cart that contains oxidizer tanks, a weighing system, a vacuum pump, a control and instrumentation panel, filters, and valves. When the unit is connected to the oxidizer hardlines and the Reaction Control Subsystem fill connections, it measures the weight of the oxidizer supplied, filters the oxidizer, monitors the oxidizer temperature, and bleeds air from the system. At the Static Test Area, this unit is remotely controlled by the RCS Stand Control Station.

5-47. OXIDIZER FLUSH AND PURGE CART.

The Oxidizer Flush and Purge Cart is a four-wheel cart that uses self-contained deionized water, triethanolamine solution, methyl alcohol, and gaseous nitrogen to provide cleaning and decontamination of the Reaction Control Subsystem tanks, systems, and components which have been exposed to the oxidizer (nitrogen tetroxide).

5-48. MOBILE FLUSH AND PURGE UNIT.

The Mobile Flush and Purge Unit is a four-wheel cart containing flushing fluid tanks, a drain tank, a fluid transfer and nitrogen purging system, and the necessary controls and instrumentation. When the unit is connected to either the Propulsion Subsystem fuel or oxidizer systems, this unit will clean, decontaminate, and dry the tankage and systems that have been exposed to either the oxidizer (nitrogen tetroxide) or the fuel (unsymmetrical dimethylhydrazine).

5-49. HELIUM PRESSURE DISTRIBUTION UNIT.

The Helium Pressure Distribution Unit is a unit containing a distribution manifold and three fill units. The fill units control and monitor helium flow to the Reaction Control Subsystem ascent and descent helium tanks when connected to the Helium Transfer Unit. For helium tank protection, the distribution unit also has relief capabilities. This unit is remotely controlled by the Propellant Control Station or by the Helium Distribution Unit Controller.

5-50. FUEL LOADING CONTROL ASSEMBLY.

The Fuel Loading Control Assembly is of modular construction and consists of a flow control system, the necessary controls and instrumentation, and provisions to isolate major components for replacement and maintenance. It is capable of being used in an unsheltered location where vibrations, noise, and explosive gases are present. When connected, the unit provides a remotely operated means of controlled propellant fuel loading, detanking, and purging of the ascent and descent fuel storage tanks. The unit is remotely controlled by the Propellant Loading Control Assembly Controller or by the Propellant Control Station.

5-51. OXIDIZER LOADING CONTROL ASSEMBLY.

The Oxidizer Loading Control Assembly is of modular construction and contains a flow control system, the necessary controls and instrumentation, and provisions to isolate major components for replacement and maintenance. It is capable of being used in an unsheltered location where vibrations, noise, and explosive gases are present. When connected, the unit provides a remotely operated means of controlled propellant oxidizer loading, detanking, and purging of the ascent and descent oxidizer storage tanks. The unit is remotely controlled by the Propellant Loading Control Assembly Controller or by the Propellant Control Station.

5-52. GASEOUS HYDROGEN SUPPLY UNIT.

The Gaseous Hydrogen Supply Unit is a four-wheel, flat, towable cart containing manifolding, gas bottles, and a control and monitoring panel. The unit will be unsheltered during operation and subject to ambient environmental conditions. When connected to the Hydrogen Transfer and Monitoring Unit, this unit will supply gaseous hydrogen at a prescribed flow and has the capability of monitoring the pressure and temperature of the gas being delivered. The unit has at least 10 standard gas bottles, which may be replaced quickly and easily, without interrupting gas output. It can be remotely controlled by the Power Generation Section Spacecraft and Servicing Remote Control Console.

5-53. GASEOUS OXYGEN SUPPLY UNIT.

The Gaseous Oxygen Supply Unit is a four-wheel, flat, towable cart containing manifolding, gas bottles, and a control and monitoring panel. The unit will be unsheltered during operation and subject to ambient environmental conditions. When connected to the Oxygen Transfer and Monitoring Unit, this unit will supply gaseous oxygen at a prescribed flow and has the capability of monitoring the pressure and temperature of the gas being delivered. The unit has at least 10 standard gas bottles, which may be replaced quickly and easily, without interrupting gas output. This unit can be remotely controlled by the Power Generation Section Spacrcraft and Servicing Remote Control Console.

5-54. REFEREE CRYOGENIC FLUID SUPPLY UNIT.

The Referee Cryogenic Fluid Supply Unit is a two-wheel tank trailer that consists of a Dewar vessel for storing 250 gallons of liquid and 50 gallons of vapor, and a fluid transfer system. This unit supplies liquid nitrogen to the Hydrogen Transfer and Monitoring Unit and to the Oxygen Transfer and Monitoring Unit. It can be remotely controlled by the Power Generation Section Spacecraft and Servicing Remote Control Console.

5-55. GASEOUS NITROGEN AND/OR HELIUM SUPPLY UNIT.

The Gaseous Nitrogen and/or Helium Supply Unit is a four-wheel cart that consists of a flat cart containing manifolding, seven nitrogen bottles, and seven helium bottles, and a panel from which monitoring and control is effected. The unit supplies gaseous nitrogen and/or helium to the Hydrogen Transfer and Monitoring Unit and to the Oxygen Transfer and Monitoring Unit. It can be remotely controlled by the Power Generation Section Spacecraft and Servicing Remote Control Console.

5-56. HYDROGEN TRANSFER AND MONITORING UNIT.

The Hydrogen Transfer and Monitoring Unit is a skid-mounted unit that contains cryogenic pumps and piping, subcoolers, conditioning equipment, instrumentation, wiring, and hardware for unit operation. The unit controls and monitors the flow of clean, dry, gaseous nitrogen and/or gaseous helium to purge the reactant storage system; and clean, dry, gaseous hydrogen to purge inert gases from the reactant storage system. It also controls and monitors filling of the reactant storage system with liquid hydrogen. This unit can be remotely controlled by the Power Generation Section Spacecraft and Servicing Remote Control Console.

5-57. OXYGEN TRANSFER AND MONITORING UNIT.

The Oxygen Transfer and Monitoring Unit is a skid-mounted unit that contains cryogenic pumps and piping, subcoolers, conditioning equipment, instrumentation, wiring, and hardware for unit operation. The unit controls and monitors the flow of clean, dry, gaseous nitrogen and/or gaseous helium to purge the liquid oxygen tanks of the reactant storage system; and clean, dry, gaseous oxygen to purge inert gases from the liquid oxygen tanks of the reactant storage system. It also controls and monitors filling the liquid oxygen tanks of the reactant storage system with clean liquid oxygen. This unit can be remotely controlled by the Power Generation Section Spacecraft and Servicing Remote Control Console.

5-58. OXIDIZER TRANSFER AND CONDITIONING UNIT.

The Oxidizer Transfer and Conditioning Unit is a four-wheel cart consisting of a temperature control section, a heat transfer section, an oxidizer section, and the necessary controls and instrumentation. When loading the ascent and descent stage oxidizer tanks with oxidizer (nitrogen tetroxide - N_2O_4), this unit controls the temperature of the oxidizer in the range from 30° to 135°F. It requires a minimum time of 2 hours to condition the 1,200 gallons of oxidizer from an ambient temperature of 40° to 80°F to either extremes of temperature. At the Static Test Area, this unit is remotely controlled by the Propellant Control Station.

5-59. WATER SUPPLY UNIT.

The Water Supply Unit is a mobile unit that contains a pump, starters, tubing, valving and controls, remote and manual controls, explosion-proofing, instrumentation, and wiring. It is used in the LEM Environmental Control Subsystem to fill the water management section with triple-distilled water, and to evacuate the water management section. This unit is an item of common usage GSE modified for LEM use.

5-60. FUEL TRANSFER AND CONDITIONING UNIT.

The Fuel Transfer and Conditioning Unit is a four-wheel cart consisting of a temperature control section, a heat transfer section, a fuel section, and the necessary controls and instrumentation. When loading the fuel tanks with propellant (unsymmetrical dimethylhydrazine - UDMH), this unit controls the temperature of the fuel being transferred. This unit is an item of common usage GSE. At the Static Test Area, this unit is remotely controlled by the Propellant Control Station.

5-61. HELIUM TRANSFER AND CONDITIONER UNIT.

The Helium Transfer and Conditioner Unit is a completely enclosed four-wheel unit for use in an outdoor, unsheltered location. It contains a gas flow control system, a heat exchanger, and an electrical system. The unit is positioned at the base of the launch tower with its inlet port connected to the helium supply and its outlet port connected to the helium distribution lines. In this configuration, conditioned gaseous helium is transferred from the storage unit to the LEM Propulsion and Reaction Control Subsystem helium tanks. This unit is remotely controlled by the Pneumatics Control Station when used at the Static Test Area, and it is an item of common usage GSE.

5-62. WATER-GLYCOL SERVICE UNIT.

The Water-Glycol Service Unit is located at the base of the Mobile Arming Tower and contains pumps, a reservoir, an accumulator, heaters, and plumbing. This unit is used to supply a glycol mixture, distilled water, and gaseous nitrogen to the Water-Glycol Trim Control Unit. It is also used to evacuate the ECS heat transport section. This cooling requirement must be met to remove heat from the LEM during prelaunch checkout. This unit is an item of common usage GSE.

5-63. HELIUM BOOSTER CART.

The Helium Booster Cart is a four-wheel cart that contains a boost pump, an electrically driven power system, inter- and after-coolers, and electrical and pneumatic controls. With the cart connected between the Helium Storage Trailer and the helium supply source, the boost pump establishes equilibrium pressure between the supply source and storage trailer and boosts the pressure of the Helium Storage Trailer. This unit is an item of common usage GSE. At the Static Test Area, the operation of this unit is monitored by the Pneumatics Control Station, which controls the power interlock emergency shutdown devices of this unit.

5-64. RCS OXIDIZER SERVICING UNIT.

The RCS Oxidizer Servicing Unit is a mobile unit that contains a holding tank, oxidizer pumping and control system, measuring system, thermal conditioning system, filters, nitrogen pressurization and an evacuation system, a control and instrumentation panel, vehicle and hardline adapter hoses, and a remote control unit. With the servicing unit connected to the RCS Oxidizer Transfer Control Unit, this unit will supply, condition, and control the oxidizer to the RCS tanks. Upon completion of the fill mode, it will unload RCS tanks, then drain and purge the RCS Oxidizer Transfer Control Unit along with the Fluid Distribution System. At the Static Test Area, this unit is remotely controlled by the RCS Stand Control Station. This servicing unit is an item of common usage GSE modified for LEM use.

5-65. FUEL READY STORAGE UNIT.

The Fuel Ready Storage Unit is a four-wheel cart that consists of a 1500 gallon storage tank and a transfer system with the necessary controls and instrumentation. With the transfer system in operation, the propellant (UDMH) is transferred to the LEM propellant tanks. This unit also can recirculate and off-load propellant from the LEM propellant tanks. This unit is used in conjunction with the Fuel Transfer Control Unit and is an item of common usage GSE. At the Static Test Area, this unit is remotely controlled by the Propellant Control Station.

5-66. OXIDIZER READY STORAGE UNIT.

The Oxidizer Ready Storage Unit is a four-wheel unit that consists of a 1500 gallon storage tank and a transfer system with the necessary controls and instrumentation. With the transfer system in operation, the oxidizer (N_2O_4) is transferred to the LEM oxidizer tanks. Also, this unit can recirculate and off-load the oxidizer from the LEM oxidizer tanks. The unit is used in conjunction with the Oxidizer Transfer Control Unit and is an item of common usage GSE. At the Static Test Area, this unit is remotely controlled by the Propellant Control Station.

5-67. FUEL TOXIC VAPOR DISPOSAL UNIT.

The Fuel Toxic Vapor Disposal Unit is a skid-mounted assembly module consisting of a gas processing system with the necessary controls and instrumentation. It is used to safely dispose of fuel vapors generated during the thermal conditioning of the fuel, during the nitrogen purging of the fuel system, and during the fuel loading operation. This unit is an item of common usage GSE. At the Static Test Area, this unit is remotely controlled by the Propellant Control Station.

5-68. OXIDIZER TOXIC VAPOR DISPOSAL UNIT.

The Oxidizer Toxic Vapor Disposal Unit is a skid-mounted assembly module consisting of a gas processing system with the necessary controls and instrumentation. It is used to safely dispose of oxidizer vapors generated during the thermal conditioning of the oxidizer, during the nitrogen purging of the system, and during the oxidizer loading operation. This unit is an item of common usage GSE. At the Static Test Area, this unit is remotely controlled by the Propellant Control Station.

5-69. HELIUM STORAGE TRAILER.

The Helium Storage Trailer is an eight-wheel, pneumatic tire, semitrailer. Mounted on the trailer are gas cylinders, which are filled from the facility helium source using the Helium Booster Cart. The LEM helium tanks are pressurized and filled from these cylinders by blowing down through the Helium Transfer and Conditioner Unit. The helium stored in the gas cylinders is sufficient to provide two fillings of the Propulsion and Reaction Control Subsystem tanks to proper pressure. This unit is an item of common usage GSE.

5-70. RCS FUEL SERVICING UNIT.

The RCS Fuel Servicing Unit is a mobile unit containing a holding tank, a fuel pumping and control system, a measuring system, a thermal conditioning system, filters, a nitrogen pressurization and evacuation system, a control and instrumentation panel, vehicle and hardline adapter hoses, and a remote control unit. With this servicing unit connected to the RCS Fuel Transfer Control Unit, this unit will supply, condition and control the fuel to the RCS tanks. Upon completion of the fill mode, it will unload the RCS tanks, then drain and purge the RCS Fuel Transfer Control Unit along with the Fluid Distribution System. At the Static Test Area, this unit is remotely controlled by the RCS Stand Control Station. This servicing unit is an item of common usage GSE modified for LEM use.

5-71. LIQUID OXYGEN MOBILE STORAGE UNIT.

The Liquid Oxygen Mobile Storage Unit is a four-wheel, pneumatic tire, semitrailer. Mounted on the trailer is a liquid oxygen storage system consisting of a Dewar storage vessel with the necessary gages, valves, and lines; a fluid transfer system; and a liquid vaporizer. Liquid and gaseous oxygen is supplied to the Oxygen Transfer and Monitoring Unit. The storage unit is capable of operating in an unsheltered area, and can be remotely controlled by the Power Generation Section Spacecraft and Servicing Remote Control Console. This unit is an item of common usage GSE.

5-72. LIQUID HYDROGEN MOBILE STORAGE UNIT.

The Liquid Hydrogen Mobile Storage Unit is a four-wheel, pneumatic tire, semitrailer. Mounted on the trailer is a liquid hydrogen storage system consisting of a Dewar storage vessel with the necessary gages, valves, and lines; a fluid transfer system; and a liquid vaporizer. Liquid and gaseous hydrogen is supplied to the Hydrogen Transfer and Monitoring Unit. The storage unit is capable of operating in an unsheltered area, and can be remotely controlled by the Power Generation Section Spacecraft and Servicing Remote Control Console. This unit is an item of common usage GSE.

5-73. CONDITION AND EQUIPMENT SIMULATORS.

5-74. AC LOAD BANK.

The AC Load Bank is a roll-about cart used to support checkout of the LEM inverter and a-c power distribution network of the Electrical Power Subsystem, whether they are installed in, or removed from the spacecraft. It supplies various a-c loads with variable power factors. Operation can be either local or remote; interfaces enable the AC Load Bank to be controlled from and monitored by: (1) the ACE-S/C, (2) the Inverter Bench Maintenance Equipment, or (3) the Distribution Bench Maintenance Equipment.

5-75. LOAD SIMULATOR.

The Load Simulator is used to test the power generation section and fuel cell assemblies of the Electrical Power Subsystem. It simulates 100 different d-c steady-state loads (as would be caused by normal operation of a subsystem) and transient loads (as would be caused by the operation of an RCS solenoid). The application of these loads are controlled by three modes of operation: (1) automatic mode, where tape program control is used to simulate the entire LEM mission d-c electrical load profile, (2) semiautomatic mode, where programmed subroutines are manually selected to simulate series of RCS-type transients, and (3) manual mode, where individual loads are manually selected.

5-76. DC TRANSIENT SUPPLY.

The DC Transient Supply is used to verify the ability of any assembly, section, or subsystem that is powered by the spacecraft d-c distribution network to operate under transient voltages in excess of those expected during the LEM mission. It supplies prime d-c power with transients at +78 volts and at -100 volts, and with a 10-microsecond pulse width. Operation can be manual or semiautomatic. If the semiautomatic mode is chosen, these transients will be applied to the equipment under test for 5 minutes at 10 pulses per second. Ripple injection is provided as a separate function, and may or may not be used simultaneously with the application of the transients.

5-77. INVERTER SUBSTITUTE UNIT.

The Inverter Substitute Unit is used to supply all spacecraft a-c bus power when it is decided against using the limited-life spacecraft inverter during prolonged testing operations. It is powered by either the LEM fuel cell assemblies or by its own internal power supply, and operates in synchronization with either the LEM pulse code modulation and timing equipment (PCMTE) synchronization signals, or with its internal synchronization generator, which simulates PCMTE synchronization signals. The simulated PCMTE signals are made available to other equipments, enabling test that require PCMTE synchronization to be performed without activating the Instrumentation Subsystem.

5-78. PORTABLE LIFE SUPPORT SYSTEM BATTERY SIMULATOR.

The Portable Life Support System (PLSS) Battery Simulator is used to verify proper operation of the spacecraft PLSS battery charger. It is a portable test set that simulates various charge and discharge states of the astronauts' backpack battery, and verifies the PLSS battery charger's response to them. All input and output currents and voltages are monitored and displayed. Interfaces enable the PLSS Battery Simulator to operate in conjunction with ACE-S/C.

5-79. HANDLING AND TRANSPORTATION EQUIPMENT AND WORKSTANDS.

5-80. AUXILIARY CRANE CONTROL.

The Auxiliary Crane Control is a self-contained, hydraulically operated unit that interconnects between the appropriate hoisting sling and the lifting device through the upper and lower eyebolt fittings. The unit consists of an upper and lower eyebolt, a return lift dial, a scale dial, up and down valve levers, and a plunger rod. There are also two control reels, each with 40 feet of nylon covered steel cable. This control operates independently of the lifting device and hoisting sling and is capable of raising and lowering loads up to five tons a distance of 12 inches with an accuracy to within 0.0001 inch or less. The cables are attached to the lever controls and afford operation of the unit from a distance.

5-81. ASCENT STAGE TRANSPORTER.

The Ascent Stage Transporter is a four-wheel trailer that consists of four main assemblies: base platform, chassis frame, suspension system, and environmental cover. The base platform forms a rectangular frame with extension frames attached to each side during ground operation. Flooring covers the structural framework. Tiedown fittings, at four points, are used for mounting and supporting the ascent stage on the base platform. The base platform is mounted to, and supported by, the chassis frame. Air brakes, electrical facilities, a tow bar, and steering linkage are mounted to the chassis. Individual suspension assemblies, tiedown provisions, and jacking facilities are attached to the main sills. The suspension system consists of four individually acting suspension assemblies. The environmental cover consists of a canopy-type, aluminum skin structure secured to the base platform. Hoisting lugs, used in conjunction with the transporter cover sling, are provided. Continuous rubber seals and tamperproof locking provisions are at the mating surface.

5-82. TRANSPORTER COVER SLING.

The Transporter Cover Sling consists of a steel spreader bar, four steel cables with a universal fitting swaged on each end, and a pear-shaped universal link attached to the center of the spreader bar. The sling is used in conjunction with a suitable crane or lifting device to lift either the ascent or descent stage transporter cover from the transporter. The sling is proof-loaded to four times the weight of the transporter cover.

5-83. ASCENT STAGE CLEAN ROOM DOLLY.

The Ascent Stage Clean Room Dolly consists of a welded, rectangular, steel tube chassis; four heavy-duty, 8-inch, swivel casters; and a tow bar secured with a quick-disconnect pin. Each caster assembly has a mechanically operated parking brake. The dolly is designed to clean room design criteria and is used to transport the ascent stage during all phases of clean room operations. It is also used for interplant transportation not requiring clean room conditions.

5-84. DESCENT STAGE TRANSPORTER.

The Descent Stage Transporter is a four-wheel trailer that consists of four main assemblies: base platform, chassis frame, suspension system, and environmental cover. The base platform forms a rectangular frame with extension frames attached to each side during ground operations. Flooring covers the structural framework. Tiedown fittings, at four points, are used for mounting and supporting the descent stage on the base platform. The base platform is mounted to, and supported by, the chassis frame. Air brakes, electrical facilities, tow bar, and steerage linkage are mounted to the chassis. Individual suspension assemblies, tiedown provisions, and jacking facilities are attached to the main skills. The suspension system consists of four individually acting suspension assemblies. The environmental cover consists of a canopy-type, aluminum skin structure secured to the base platform. Hoisting lugs, used in conjunction with the transporter cover sling, are provided. Continuous rubber seals and tamperproof locking provisions are at the mating surface.

5-85. DESCENT STAGE CLEAN ROOM DOLLY.

The Descent Stage Clean Room Dolly consists of a welded, rectangular steel tube chassis; four heavy-duty, 8-inch swivel casters; and a tow bar secured with a quick-disconnect pin. Each caster assembly has a mechanically operated parking brake. The dolly is designed to clean room design criteria and is used to transport the descent stage or the complete LEM (less landing gear) during all phases of clean room operation. It is also used for interplant transportation not requiring clean room conditions.

5-86. DESCENT STAGE HARDMOUNT ADAPTER SET.

The Descent Stage Hardmount Adapter Set consists of four builtup structural steel adapters and is mounted to the actuation subsystem supports. The adapters have fittings which pick up the structural hard points of the descent stage. This unit is capable of supporting the descent stage or the complete LEM (less landing gear) during static firing tests. The adapters include an electrically nonconductive section so that the test vehicle is not electrically grounded to the test chamber. With the descent stage mounted on the adapter set, the descent stage engine may be fired and gimballed 6 degrees.

5-87. ASCENT STAGE HARDMOUNT ADAPTER.

The Ascent Stage Hardmount Adapter is constructed from structural steel welded to form a rigid frame. Lifting eyes and a cable sling are included for ease of installation in the static test firing chamber. The adapter is mounted on the actuation subsystem supports. An open section, in the center of the adapter, is provided so that the ascent stage engine may be removed when the ascent stage is raised by the actuation subsystem. There is interface with the ascent stage soft mounts.

5-88. MOBILE CRANE.

The Mobile Crane is a motorized boom crane capable of lifting the ascent and/or descent stage from the cargo lift trailer to the ascent and/or descent stage transporter. This unit is an item of GFE (government furnished equipment).

5-89. ASCENT STAGE WORKSTAND.

The Ascent Stage Workstand is a two-level, knock-down-type stand capable of supporting 18 men with tools and parts. It consists of vertical trusses that support an aluminum, diamond-plate flooring. The upper partial platform provides access to the top docking hatch. The workstand, comprised of two sections, is assembled to envelop the ascent stage and provide 360 degree access and separates for the removal of the stage. Each section has retractable casters that permit the section to be rolled, by hand or tow tractor, for removal or assembly. Kickplates at floor level prevent tools or equipment from inadvertently falling off. Railings provide for personnel safety. Stairways provide access to both levels of the workstand.

5-90. DESCENT STAGE WORKSTAND.

The Descent Stage Workstand is a two-level, knock-down-type stand capable of supporting 18 men with tools and parts. It consists of vertical trusses that support an aluminum, diamond-plate flooring. The workstand, comprised of two sections, is assembled to envelop the descent stage and provide 360 degrees access and separates for the removal of the stage. Each section has retractable casters to permit the section to be rolled, by hand or tow tractor, for removal or assembly. Kickplates at the floor level prevent tools or equipment from inadvertently falling off. Railings provide for personnel safety. Stairways provide access to both levels of the workstand.

5-91. DESCENT STAGE SUPPORT STAND.

The Descent Stage Support Stand is a structural steel stand used to support and position the descent stage. It is capable of supporting the entire LEM and permits mating the ascent and descent stages. The descent stage is positioned high enough to permit the installation of the landing gear, with the gear above floor level. The stand has a steel frame with support and securing provisions that mate with the descent stage hard points. Four vertical columns maintain the LEM at the proper working height. Each column has a bearing plate with provisions for cinching the stand to the floor.

5-92. CLEANING POSITIONER.

The Cleaning Positioner is a unit consisting of a gimballed platform supported by two floor-mounted stanchions; two electrical motors, each with a reduction gear unit and a chain drive unit; and a remote control console. A rectangular turntable is mounted on the platform. Both the platform and turntable can rotate 360 degrees at a speed of one rpm. This speed is achieved by each of the electric motors driving its reduction gear and chain drive unit. The Cleaning Positioner is used to rotate either the ascent or descent stage about the pitch and roll axes, thereby enabling the interior of the stage to be vacuum-cleaned, and thus precluding the possibility of foreign matter remaining within the stage.

5-93. BENCH MAINTENANCE EQUIPMENT.

Bench Maintenance Equipment is presented in Tables 5-1 through 5-6 according to the subsystem to which the equipment applies.

Table 5-1. Navigation and Guidance Subsystem

Nomenclature	Used to check and troubleshoot:
Landing radar and rendezvous radar/transponder bench maintenance equipment	Landing radar and rendezvous radar/transponder to replaceable component level; simulates analog and digital inputs; checks range and altitude measuring capabilities; measures range signal stability and longtime frequency stability of oscillator sections; records outputs; and displays spectral waveforms for periodic maintenance, overhaul, and calibration.

Table 5-2. Stabilization and Control Subsystem

Nomenclature	Used to check and troubleshoot:
Abort guidance section bench maintenance equipment	The complete abort guidance section or its assemblies, which are the sensor assembly, electronics assembly, and abort programmer assembly to the replaceable component level.
Control electronics section bench maintenance equipment	The complete control electronics section or its assemblies, which are the rate gyro assembly, descent engine control assembly, attitude controller, translation controller, and gimbal drive actuator assembly and the complete in-flight monitor assembly to the replaceable component level.

Table 5-3. Electrical Power Subsystem

Nomenclature	Used to check and troubleshoot:
Distribution bench maintenance equipment	Power distribution section to replaceable component level; measures input and output voltage and current, continuity and logic, response times, and transients.
Inverter bench maintenance equipment	Inverter assembly to replaceable subassembly level by applying variable power factor a-c loads and variable d-c voltage inputs, while monitoring frequency regulation, voltage regulation, transient response, harmonic distortion, and critical waveforms.
Battery charger bench maintenance equipment	Portable life support system (PLSS) battery charger to replaceable subassembly level by simulating all possible discharge states of the PLSS battery and monitoring the PLSS battery charger's response to these conditions.
Electrical lighting power and control bench maintenance equipment	Power controls of the LEM electric lighting system by providing simulated lighting loads while monitoring control functions.
Pyrotechnic bench maintenance equipment	Pyrotechnic distribution circuits and subassemblies by simulating pyrotechnic device while controlling and monitoring pyrotechnic sequencing and firing logic. Can be used with Pyrotechnic Initiator Test Set, enabling sequencing, firing logic, and squib valve ignitor bridgewire verification.

Table 5-4. Communications Subsystem

Nomenclature	Used to check and troubleshoot:
Antenna bench maintenance equipment	UHF and VHF antennas and IR sensor, and electronics and gimbal system of steerable antenna; checks VSWR, chopper-filter, d-c voltage versus angle-off-boresight, antenna overshoot, linearity, and threshold voltage.
S-band transponder bench maintenance equipment	S-band transponder to replaceable component level; checks input power, frequency stability, noise figure, image response, linearity, receiver threshold, phase error, bandwidth, phase modulation, frequency modulation, and power output.
C-band transponder bench maintenance equipment	C-band transponder to replaceable component level; checks input power, frequency stability, noise figure, image response, linearity, receiver threshold, phase error, bandwidth, and power output.
Communications equipment bench maintenance equipment	Transmitters, receivers, and antennas of Communications Subsystem to replaceable component level.
TV camera bench maintenance equipment*	TV camera to replaceable component level; checks video preamplifier output, d-c voltage regulator, and composite video circuitry; measures grid voltage, horizontal and vertical sweep, horizontal and vertical blanking trigger, and power input.

* Used on both LEM and Command/Service Modules (common usage).

Table 5-5. Controls and Displays

Nomenclature	Used to check and troubleshoot:
Ball attitude indicator bench maintenance equipment	Ball attitude indicator to replaceable component level.
Flight control bench maintenance equipment	Flight control, reaction control, and environmental control panels to replaceable component level.
Power and radar bench maintenance equipment	Radar, power generation, and power distribution panels to replaceable component level.
Stabilization - navigation and guidance bench maintenance equipment	Stabilization and control, and main propulsion (ascent/descent) panels to replaceable component level.

Table 5-5. Controls and Displays (cont)

Nomenclature	Used to check and troubleshoot:
Communications panel bench maintenance equipment	Communications and audio control panels to replaceable component level.

Table 5-6. Instrumentation Subsystem

Nomenclature	Used to check and troubleshoot:
Spacecraft Instrumentation Test Equipment (SITE)*	Instrumentation Subsystem and ACE-S/C Down-Link. SITE enables checkout of an entire Instrumentation Subsystem or ACE-S/C Down-Link; a section (grouping of interrelated assemblies) of either of these equipments, or individual assemblies of either of these equipments.

* Used on both LEM and Command/Service Modules (common usage)

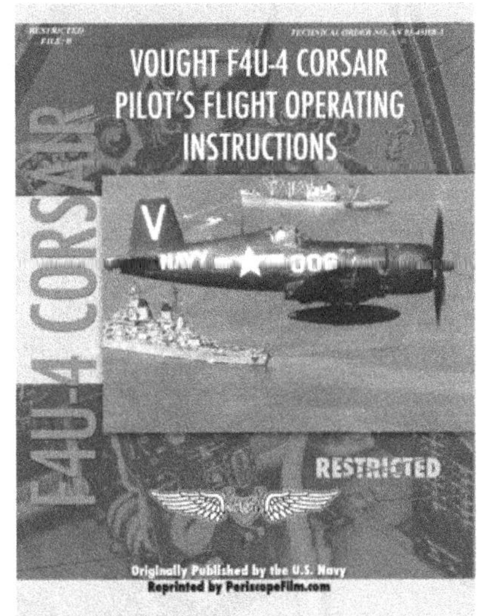
VOUGHT F4U-4 CORSAIR
PILOT'S FLIGHT OPERATING
INSTRUCTIONS
F4U-4 CORSAIR
RESTRICTED
Originally Published by the U.S. Navy
Reprinted by PeriscopeFilm.com

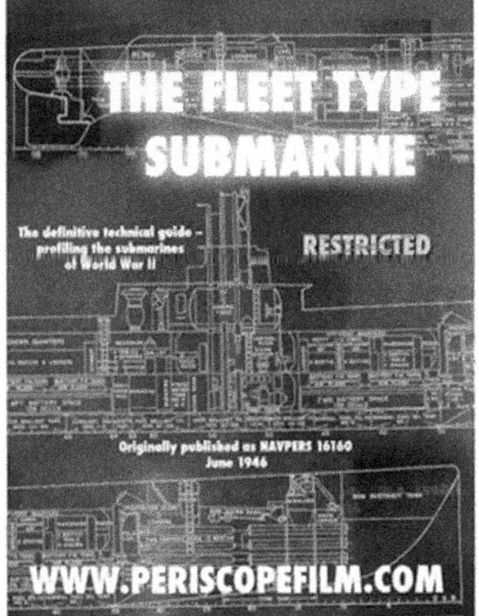
THE FLEET TYPE
SUBMARINE
The definitive technical guide - profiling the submarines of World War II
RESTRICTED
Originally published as NAVPERS 16160
June 1946
WWW.PERISCOPEFILM.COM

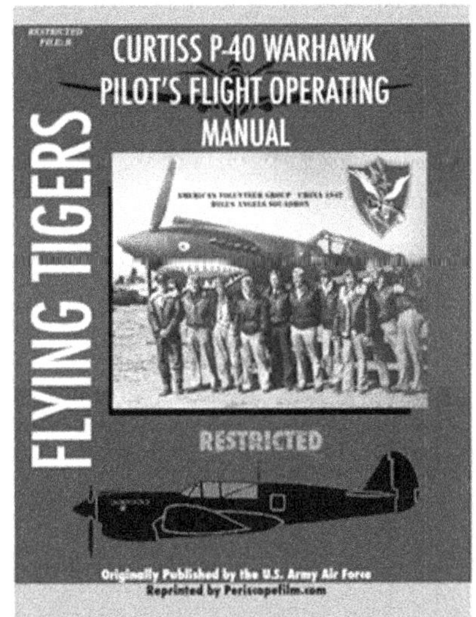
CURTISS P-40 WARHAWK
PILOT'S FLIGHT OPERATING
MANUAL
FLYING TIGERS
RESTRICTED
Originally Published by the U.S. Army Air Force
Reprinted by Periscopefilm.com

NORTH AMERICAN X-15 ROCKET PLANE
PILOT'S FLIGHT OPERATING
INSTRUCTIONS
X-15 ROCKET PLANE
RESTRICTED
X-15
Originally Published by North American Aviation and the USAF
Reprinted by Periscopefilm.com

NORTHROP P-61 BLACK WIDOW
PILOT'S FLIGHT OPERATING
INSTRUCTIONS
NIGHT FIGHTER
RESTRICTED
Originally Published by USAAF July 1945
Reprinted by Periscopefilm.com

NORTHROP YB-49 FLYING WING
PILOT'S FLIGHT OPERATING
INSTRUCTIONS
FLYING WING
RESTRICTED
Reprinted by Periscopefilm.com

NORTH AMERICAN P-51 MUSTANG
PILOT'S FLIGHT OPERATING
INSTRUCTIONS
P-51 MUSTANG
RESTRICTED

B-17F
AIRPLANE
PILOT'S FLIGHT OPERATING
INSTRUCTIONS
B-17F
Originally published by the
United States Army Air Forces
December, 1942.
RESTRICTED
Reprinted by Periscopefilm.com

LOCKHEED SR-71 BLACKBIRD
PILOT'S FLIGHT OPERATING
INSTRUCTIONS
SR-71 BLACKBIRD
RESTRICTED
Originally Published by Lockheed and the USAF

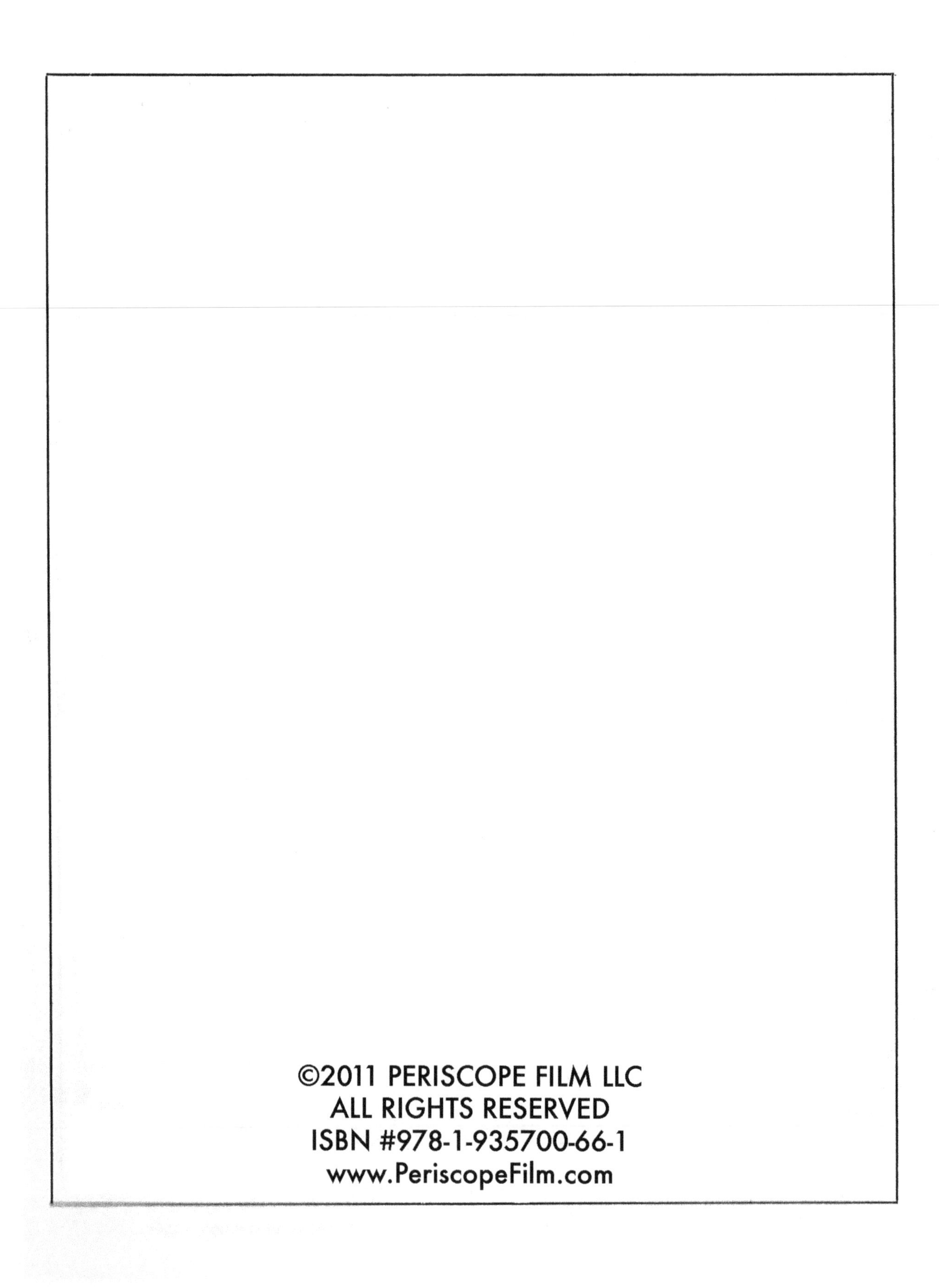

ISBN #978-1-935700-66-1
www.PeriscopeFilm.com

Lightning Source UK Ltd.
Milton Keynes UK
UKHW032008050819
347457UK00002B/10/P

9 781935 700661